1990

MATHEMATICAL MODELLING

Mathematical Modelling

J. N. Kapur

JOHN WILEY & SONS

New York   Chichester   Brisbane   Toronto   Singapore

First Published in 1988 by
WILEY EASTERN LIMITED
4835/24 Ansari Road, Daryaganj
New Delhi 110 002, India

Distributors:

Australia and New Zealand
JACARANDA WILEY LTD., JACARANDA PRESS
JOHN WILEY & SONS, INC.
GPO Box 859, Brisbane, Queensland 4001, Australia

Canada:
JOHN WILEY & SONS CANADA LIMITED
22 Worcester Road, Rexdale, Ontario, Canada

Europe and Africa:
JOHN WILEY & SONS LIMITED
Baffins Lane, Chichester, West Sussex, England

South East Asia:
JOHN WILEY & SONS, INC.
05-05, Block B, Union Industrial Building
37 Jalan Pemimpin, Singapore 2057

Africa and South Asia:
WILEY EASTERN LIMITED
4835/24 Ansari Road, Daryaganj
New Delhi 110 002, India

North and South America and rest of the world:
JOHN WILEY & SONS, INC.
605 Third Avenue, New York, NY 10158 USA

Library of Congress Cataloging-in-Publication Data

Kapur, Jagat Narain, 1923 -
        Mathematical modelling.

        Bibliography : p.
        1. Mathematical models. I. Title.
QA401.K283   1988 511'.8   85-22734

ISBN 0-470-20088-X John Wiley & Sons, Inc.
ISBN 81-224-0006-X Wiley Eastern Limited

Printed in India at Taj Press. Delhi 110006

Preface

Though Mathematical Modelling has been successfully used by almost all scientists and engineers throughout the ages, its importance as a discipline to be studied and cultivated, has been realised only during the last three or four decades. The evidence for this is provided by the six International Congresses on Mathematical Modelling and three International Conferences on Teaching of Mathematical Modelling that have been held during this period as well as by about a dozen journals on Teaching and Research on Mathematical Modelling that have been started during the last two decades.

A large number of excellent text-books on Mathematical Modelling have also appeared. These are either discipline-centred or technique-centred or situations-centred.

In the first category, we have books on Mathematical Physics, Mathematical Biology, Mathematical Economics, Mathematical Sociology, Mathematical Psychology as well as books on Mathematical Models in Population Dynamics or in Biomechanics or in Water Resources or in Pollution Control etc. The emphasis in these books is not primarily on mathematical modelling; it is rather on getting an insight into the discipline concerned through the use of mathematical techniques. All mathematical-techniques are used, but these are applied to problems in one discipline only.

In the second category, we have books on Mathematical Modelling through a special class of mathematical techniques. These techniques may belong to linear algebra, ordinary differential equations, partial differential equations, difference equations, integral equations, graph theory, calculus of variations, linear and non-linear programming, dynamic programming etc. Here mathematical models are considered from different disciplines, but the choice is restricted to those models which can be understood through the particular class of techniques.

In the third category, we have books on case-studies of Mathematical Models. Various special situations are considered and mathematical models are developed for them. The situations may arise in any discipline and may be even interdisciplinary. The model may require one or more mathematical techniques for its solution. The object is to get insight into the situation concerned. It is expected that students will be able to transfer the learning gained from special case-studies to other situations.

Though some mathematical modelling problems may require more than one technique for their solution, yet in practice most mathematical modelling problems, use one dominant technique. It will be accordingly useful to know as to which technique is most appropriate for a particular situation. The present book aims to provide some help in answering the question.

The first chapter explains the basic principles of mathematical modelling and illustrates these with some simple but very important examples of mathematical modelling through geometry, algebra, trigonometry and calculus.

Each subsequent chapter deals with mathematical modelling through one or more specific techniques. Thus we consider mathematical modelling through ordinary differential equations of first and second order, through systems of ordinary differential equations, through difference equations, through functional equations, through integral, integro-differential, differential-difference, delay-differential and partial differential equations, through graph theory concepts, through linear and non-linear programming, through dynamical programming, through calculus of variations, through maximum principle and through maximum entropy principle.

In each chapter, mathematical models are chosen from physical, biological, social, economic, management and engineering sciences. The models deal with different concepts, but have a common mathematical structure and bring out the unifying influence of mathematical modelling in different disciplines. Thus quite different problems in Physics, Biology, Economics, Psychology and Engineering may be represented by a common mathematical model. Here model is the same, only the interpretations are different. Efforts are made to explicitly bring out when different techniques are most appropriate.

Mathematical Modelling can be learnt by making mathematical models. With this object, plenty of exercises have been specially constructed. These are of three types. Some of these develop further the models given in the text; others ask students to change the hypotheses of the models given in the text and to deduce the consequences of the new models and compare with the consequences of the earlier models; still others ask the students to develop models for completely new situations.

In most books used in mathematics courses, the emphasis is on techniques and the emphasis on mathematical modelling is given a secondary place. In the present book, the emphasis is reversed. Here mathematical modelling is of primary importance and mathematical techniques are of secondary importance.

The book can be used at senior undergraduate or graduate level. About two-thirds of the book can be covered in one semester and the whole book can be covered in depth in a two-semester course.

The book not only aims to introduce the student to mathematical modelling, it also aims to give the students a panaromic view of applications of mathematics in science and technology and thus it aims to correct some of the imbalance which occurs in some curricula where applications are not sufficiently emphasised and which give students only a partial and incomplete picture of Mathematics.

I have given courses on Mathematical Modelling at I.I.T. Kanpur, Manitoba University and University of Waterloo to students from mathematics, science, engineering and commerce departments. I have also given special courses on Mathematical Models in Biology and Medicine and Maximum Entropy Models in Science and Engineering based on my books on these subjects. I have also given more than a hundred seminars on various aspects of Mathematical Modelling at more than fifty universities in India, U.K., U.S.A., Canada, Italy and Australia. I am grateful to all my students and to all scientists from all over the world with whom I had the privilege of useful discussions on Mathematical Modelling.

I am also grateful to my publishers for their patience and help at all stages of the writing of this book. I am also grateful to Quality Improvement Programme at IIT Kanpur for financial support for writing this book.

This book is dedicated to Mathematical Sciences Trust Society to strengthen its efforts for the development of Mathematical Sciences in India.

J.N. KAPUR

Contents

Preface   *iii*

CHAPTER 1. **Mathematical Modelling: Need, Techniques,**
            Classifications and Simple Illustrations   **1**

    1.1 Simple Situations Requiring Mathematical Modelling *1*
    1.2 The Technique of Mathematical Modelling *5*
    1.3 Classification of Mathematical Models *7*
    1.4 Some Characteristics of Mathematical Models *9*
    1.5 Mathematical Modelling Through Geometry *14*
    1.6 Mathematical Modelling Through Algebra *16*
    1.7 Mathematical Modelling Through Trigonometry *20*
    1.8 Mathematical Modelling Through Calculus *23*
    1.9 Limitations of Mathematical Modelling *28*

CHAPTER 2. **Mathematical Modelling Through Ordinary Differential**
            Equations of First Order   **30**

    2.1 Mathematical Modelling Through Differential
             Equations *30*
    2.2 Linear Growth and Decay Models *30*
    2.3 Non-Linear Growth and Decay Models *35*
    2.4 Compartment Models *39*
    2.5 Mathematical Modelling in Dynamics Through
             Ordinary Differential Equations of First Order *43*
    2.6 Mathematical Modelling of Geometrical Problems
             Through Ordinary Differential Equations of First
             Order *48*

CHAPTER 3. **Mathematical Modelling Through Systems of Ordinary**
            Differential Equations of the First Order   **53**

    3.1 Mathematical Modelling in Population Dynamics *53*
    3.2 Mathematical Modelling of Epidemics Through Systems
             of Ordinary Differential Equations of First Order *60*
    3.3 Compartment Models Through Systems of Ordinary
             Differential Equations *63*
   ·3.4 Mathematical Modelling in Economics Through
             Systems of Ordinary Differential Equations of
             First Order *64*

3.5 Mathematical Models in Medicine, Arms Race, Battles and International Trade in Terms of Systems of Ordinary Differential Equations *69*

3.6 Mathematical Modelling in Dynamics Through Systems of Ordinary Differential Equations of First Order *72*

CHAPTER 4. **Mathematical Modelling Through Ordinary Differential Equations of Second Order** **76**

4.1 Mathematical Modelling of Planetary Motions *76*

4.2 Mathematical Modelling of Circular Motion and Motion of Satellites *82*

4.3 Mathematical Modelling Through Linear Differential Equations of Second Order *88*

4.4 Miscellaneous Mathematical Models Through Ordinary Differential Equations of the Second Order *93*

CHAPTER 5. **Mathematical Modelling Through Difference Equations** **96**

5.1 The Need for Mathematical Modelling Through Difference Equations: Some Simple Models *96*

5.2 Basic Theory of Linear Difference Equations with Constant Coefficients *98*

5.3 Mathematical Modelling Through Difference Equations in Economics and Finance *105*

5.4 Mathematical Modelling Through Difference Equations in Population Dynamics and Genetics *110*

5.5 Mathematical Modelling Through Difference Equations in Probability Theory *117*

5.6 Miscellaneous Examples of Mathematical Modelling Through Difference Equations *121*

CHAPTER 6. **Mathematical Modelling Through Partial Differential Equations** **124**

6.1 Situations Giving Rise to Partial Differential Equations Models *124*

6.2 Mass-Balance Equations: First Method of Getting PDE Models *126*

6.3 Momentum-Balance Equations: The Second Method of Obtaining Partial Differential Equation Models *132*

6.4 Variational Principles: Third Method of Obtaining Partial Differential Equation Models *136*

6.5 Probability Generating Function, Fourth Method of Obtaining Partial Differential Equation Models *139*

6.6 Model for Traffic Flow on a Highway *141*

6.7 Nature of Partial Differential Equations *145*

6.8 Initial and Boundary Conditions *147*

CHAPTER 7. **Mathematical Modelling Through Graphs**   **151**

7.1   Situations that can be Modelled Through Graphs   *151*
7.2   Mathematical Models in Terms of Directed Graphs   *154*
7.3   Mathematical Models in Terms of Signed Graphs   *161*
7.4   Mathematical Modelling in Terms of Weighted
     Digraphs   *164*
7.5   Mathematical Modelling in Terms of Unoriented
     Graphs   *170*

CHAPTER 8. **Mathematical Modelling Through Functional Integral,
Delay-Differential and Differential-Difference Equations**   **177**

8.1   Mathematical Modelling Through Functional
     Equations   *177*
8.2   Mathematical Modelling Through Integral Equations   *184*
8.3   Mathematical Modelling Through Delay-Differential
     and Differential-Difference Equations   *194*

CHAPTER 9. **Mathematical Modelling Through Calculus of
Variations and Dynamic Programming**   **201**

9.1   Optimization Principles and Techniques   *201*
9.2   Mathematical Modelling Through Calculus of
     Variations   *205*
9.3   Mathematical Modelling Through Dynamic
     Programming   *215*

CHAPTER 10. **Mathematical Modelling Through Mathematical
Programming, Maximum Principle and Maximum-
Entropy Principle**   **225**

10.1   Mathematical Modelling Through Linear
      Programming   *225*
10.2   Mathematical Modelling Through Non-Linear
      Programming   *232*
10.3   Mathematical Modelling Through Maximum
      Principle   *237*
10.4   Mathematical Modelling Through the Use of
      Principle of Maximum Entropy   *240*

APPENDIX I: Mathematical Models Discussed in the Book   **249**

APPENDIX II: Supplementary Biblioghaphy   **253**

INDEX   **257**

1

Mathematical Modelling: Need, Techniques, Classifications and Simple Illustrations

1.1 SIMPLE SITUATIONS REQUIRING MATHEMATICAL MODELLING

Consider the following problems:

(i) Find the height of a tower, say the Kutab Minar at New Delhi or the leaning tower at Pisa (without climbing it!)

(ii) Find the width of a river or a canal (without crossing it!)

(iii) Find the mass of the Earth (without using a balance!)

(iv) Find the temperature at the surface or at the centre of the Sun (without taking a thermometer there!)

(v) Estimate the yield of wheat in India from the standing crop (without cutting and weighing the whole of it!)

(vi) Find the volume of blood inside the body of a person (without bleeding him to death!)

(vii) Estimate the population of China in the year 2000 A.D. (without waiting till then!)

(viii) Find the time it takes a satellite at a height of 10,000 kms, above the Earth's surface to complete one orbit (without sending such a satellite into orbit!)

(ix) Find the effect on the economy of 30 per cent reduction in income-tax (without actually reducing the rate!)

(x) Find the gun with the best performance when the performance depends on ten parameters, each of which can take 10 values (without manufacturing 10^{10} guns!)

(xi) Estimate the average life span of a light bulb manufactured in a factory (without lighting each bulb till it gets fused!)

(xii) Estimate the total amount of insurance claims a company has to pay next year (without waiting till the end of that year!)

All these problems and thousands of similar problems can be and have been solved through mathematical modelling.

One technique of solving the above problems is similar to that of solving 'word problems' in algebra. Suppose the age of a father is four times the age of his son and we are told that after five years, the age of the father will be only three times the age of the son. We have to find their ages. Let x be the age of the father and y be the age of the son, then the data of the problem gives

$$x = 4y, \qquad x + 5 = 3(y + 5), \tag{1}$$

giving $x = 40$, $y = 10$. The two equations of (1) give a mathematical model of the biological situation, so that the biological problem of ages is reduced to the mathematical problem of the solution of a system of two algebraic equations. The solution of the equations is finally interpreted biologically to give the ages of the father and the son.

In the same way to solve a given physical, biological or social problem, we first develop a mathematical model for it, then solve the model and finally interpret the solution in terms of the original problem.

One principle of great importance to science is the following. Whenever we want to find the value of an entity which cannot be measured directly, we introduce symbols $x, y, z, \ldots$ to represent the entity and some others which vary with it, then we appeal to laws of physics, chemistry, biology or economics and use whatever information is available to us to get relations between these variables, some of which can be measured or are known and others which cannot be directly measured and have to be found out. We use the mathematical relations developed to solve for the entities which cannot be measured directly in terms of those entities whose values can be measured or are known.

The mathematical, relations we get may be in terms of algebraic, transcendental, differential, difference, integral, integro-differential, differential-difference equations or even in terms of inequalities. Thus

For (i), we try to express the height of the tower in terms of some distances and angles which can be measured on the ground,

For (ii), we try to express the width of the river in terms of some distances and angles which can be measured on our side of the river,

For (iii), we try to express the mass of the Earth in terms of some known masses and distances,

For (iv), we try to express the temperatures at the surface and the centre of the Sun in terms of the properties of light received from its surface,

For (v), we try to find the area under wheat and the average yield per acre by cutting and weighing the crop from some representative plots,

For (vi), we inject some glucose into the blood stream and find the increase in the concentration of sugar in blood,

For (vii), we extrapolate from data from previous censuses or develop a model expressing the population as a function of time,

For (viii), we try to use Newton's laws to get a relation between the orbital period and the height of the satellite above the surface of the Earth,

For (ix), we examine the effects of similar cuts in the past or develop a mathematical model giving relation between income-tax cuts, purchasing power in the hands of individuals and its effects on productivity and inflation etc.

For (x), we develop a theory of internal ballistics of guns based on laws of burning of propellants, motion of gases inside a gun and motion of the shot inside it.

For (xi), we take a random sample of bulbs, find their life-span and use statistical inference models to estimate the life span for the population of bulbs.

For (xii), we use probabilistic models for life expectancy of individuals.

Now consider specifically problem (vi). In order to express population as a function of time, we need some hypotheses. Let us postulate that the increase in population in a unit time is equal to the excess of births in that time over the number of deaths in that time and the number of births and deaths are proportional to the size of the population. There hypotheses will give us a mathematical model whose solution gives population size as a function of time. We compare the predictions of the model with the sizes of the actual population in the past. If the agreement is good and no significant changes are taking place in birth and death rates, we can use the predictions of the model for estimating future populations. If on the other hand, the agreement is not good, we modify the hypotheses in the light of the discrepancies noted and go on modifying the hypotheses till we get good agreement between observations and predictions.

Now consider problem (x). Here the hypotheses have to come from laws of physics, chemistry and gas dynamics. We require laws of conservation of energy and momentum, laws concerning burning of propellants and laws concerning motion of gases produced. Moreover these laws have to be translated mathematically and expressed in terms of differential equations to give a mathematical model whose predictions will have to be compared with observations. Once we have the right model, no further experimentation will be necessary and the optimum gun can be found theoretically.

The above discussion explains to some extent what mathematical modelling is and why it is so useful. Instead of dealing with a tower or a river or a human body, we have to deal with mathematical equations on paper. Of course we still need some measurements, but these are kept to a minimum and mathematical modelling itself may suggest the most appropriate measurements needed.

It is much easier to solve the mathematical equations, provided we know how to formulate them and now to solve them! It is also much cheaper. Moreover quite often, it is the only way to solve problems. Thus in measuring volume of blood in the human body or mass of Earth or temperature of the Sun or life-span of a light bulb, the direct methods are impossible to use and mathematical modelling is the only alternative.

EXERCISE 1.1

1. The angles of elevation of the foot and the top of a flag-post on a tower, from a point *a* metres from the foot tower are α and β respectively (Figure 1.1). Show that the height of the flag-post is

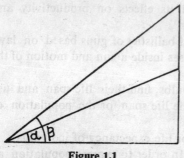

a(tan β − tan α) metres

Figure 1.1

2. Explain how you would find the breadth of a river without crossing it (use Figure 1.2).

3. You have to dig a tunnel through a mountain from *A* to *B* (Figure 1.3). Find the angle θ in terms of distances which can be measured:

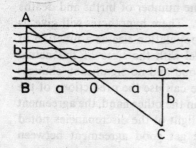

Figure 1.2

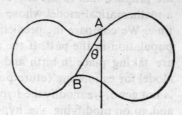

Figure 1.3

4. Show that of all rectangles with a given perimeter, the square has the maximum area. Show also that of all ractangles with a given area, the square has the minimum perimeter.

5. Let *A* and *B* be two places *d* miles apart on the surface of the Earth and having the same longitude and with latitudes θ_1° and θ_2° respectively. Show that the radius of the Earth in miles is given by

$$a = d\,\frac{180}{\pi(\theta_2 - \theta_1)}$$

6. 5 mgs of glucose are introduced into the blood stream and after 2 minutes, a samples of 10 c.c. of blood is taken in which the increase in blood sugar is found to be 0.01 mg. Estimate the volume of blood in the body.

7. A random sample of 100 light bulbs is found to have a mean life span of 200 hours and a S.D. of 10 hours. What statement, can you make about bulbs made in the factory?

8. Explain how you would find the volume of water in a village pond.

9. Suggest some methods of estimating the heights of mountain peaks and depths of ocean beds.

10. Discuss the mathematical bases of the methods used by civil engineers in land surveys.

1.2 THE TECHNIQUE OF MATHEMATICAL MODELLING

Mathematical modelling essentially consists of translating real world problems into mathematical problems, solving the mathematical problems and interpreting these solutions in the language of the real world (Figure 1.4).

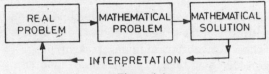

Figure 1.4

This is expressed figuratively by saying that we catch hold of the real world problem in our teeth, dive into the mathematical ocean, swim there for some time and we come out to the surface with the solution of the real world problem with us. Alternatively we may say that we soar high into the mathematical atmosphere along with the problem, fly there for some time and come down to the earth with the solution.

A real world problem, in all its generality can seldom be translated into a mathematical problem and even if it can be so translated, it may not be possible to solve the resulting mathematical problem. As such it is quite often necessary to 'idealise' or 'simplify' the problem or approximate it by another problem which is quite close to the original problem and yet it can be translated and solved mathematically. In this idealisation, we try to retain all the essential features of the problem, giving up those features which are not very essential or relevant to the situation we are investigating.

Sometimes the idealisation assumptions may look quite drastic. Thus for considering the motions of planets, we may consider the planets and Sun as point masses and neglect their sizes and structures. Similarly for considering the motion of a fluid, we may treat it as a continuous medium and neglect its discreate nature in terms of its molecular structure. The justification for such assumptions is often to be found in terms of the closeness of the agreement between observations and predictions of the mathematical models.

This leads us to modify Figure 1.4 to the following Figure 1.5.

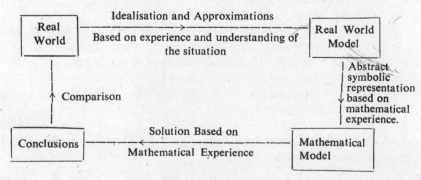

Figure 1.5

If the comparison is not satisfactory, we modify either the idealisation assumptions or search for another structure for the mathematical model.

This leads to the following twelve-point procedure for solving problems through mathematical modelling:

(i) Be clear about the real world situation to be investigated. Find all its essential characteristics relevant to the situation and find those aspects which are irrelevant or whose relevance is minimal. It is important to decide what aspects must be considered and what aspects can be ignored.

(ii) Think about all the physical, chemical, biological, social, economic laws that may be relevant to the situation. If necessary collect some data and analyse it to get some initial insight into this situation.

(iii) Formulate the problem in Problem Language (PL).

(iv) Think about all the variables $x_1, x_2, \ldots, x_n$ and parameters

$$a_1, a_2, \ldots, a_m$$

involved. Classify these into known and unknown ones.

(v) Think of the most appropriate mathematical model and translate the problem suitably into mathematical language (ML) in the form

$$f_j\left(x_i, a_h, \frac{\partial}{\partial x_l}, \int \ldots dx_i, d\right) \leqslant 0 \qquad (2)$$

i.e. in terms of algebraic, transcendental, differential, difference, integral, integro-differential, differential-difference equations or inequations.

(vi) Think of all possible ways of solving the equations of the model. The methods may be analytical, numerical or simulation. Try to get as far as possible analytically, supplement this with numerical and computer methods when necessary and use simulation when warranted.

(vii) If a reasonable change in the assumptions makes analytical solution possible, investigate the possibility. If new methods are required to solve the equations of the model, try to develop these methods.

(viii) Make an error analysis of the method used. If the error is not within acceptable limits, change the method of solution.

(ix) Translate the final solution into P.L.

(x) Compare the predictions with available observations or data. If agreement is good, accept the model. If the agreement is not good, examine the assumptions and approximations and change them in the light of the discrepancies observed and proceed as before.

(xi) Continue the process till a satisfactory model is obtained which explains all earlier data and observations.

(xii) Deduce conclusions from your model and test these conclusions against earlier data and additional data that may be collected and see if the agreement still continues to be good.

This technique will be repeatedly used in the book. However, since most of the models considered are already well-established models, their validation will not be considered, but for modelling of new situations, validation is essential.

EXERCISE 1.2
Develop mathematical models for the following situations. In each case indicate the data you would require:

1. Building a cycle stand for keeping 100 cycles in a college compound.
2. Building a play-ground complex for cricket, hockey, football and tennis.
3. Estimating the population of fish in a pond.
4. Estimating the population of lions in a forest.
5. Estimating the number of unemployed persons in a city.
6. Estimating the pollution of water of river Ganges at various points in its course.
7. Locating a common play-ground for a number of colleges.
8. Finding optimum time of replacement of machines in a factory.
9. Determining the inventory of various items a shopkeeper should keep in his shop.
10. Predicting the enrolments in various subjects in a university next year.
11. Predicting the number of mathematical modelling specialists required in the next five years.
12. Estimating the level of economic development in India so that it may catch up with the USA by the year 2000 A.D.

1.3 CLASSIFICATION OF MATHEMATICAL MODELS

(a) Mathematical Models (M.M.) may be classified according to the *subject matter* of the models. Thus we have M.M. in Physics (Mathematical Physics) M.M. in Chemistry (Theoretical Chemistry); M.M. in Biology (Mathematical Biology), M.M. in Medicine (Mathematical Medicine), M.M. in Economics (Mathematical Economics and Econometrics), M.M. in Psychology (Mathematical Psychology), M.M. in Sociology (Mathematical Sociology), M.M. in Engineering (Mathematical Engineering) and so on.

We have similarly M.M. of transportation, of urban and regional planning, of pollution, of environment, of oceanography, of blood flows, of genetics, of water resources, of optimal utilization of exhaustible and renewable resources, of political systems, of land distribution, of linguistics and so on.

In fact every branch of knowledge has two aspects, one of which is theoretical, mathematical, statistical and computer-based and the other of which is empirical, experimental and observational. Mathematical Modelling is essential to the first of these two aspects.

We have separate books on mathematical models in each of the areas we have mentioned above and in many others. One can spend a life-time specialising in mathematical models in one specified area alone.

(b) We may also classify Mathematical Models according to the *mathematical techniques* used in solving them. Thus we have Mathematical Modelling (M.M.) through classical algebra, M.M. through linear algebra and matrices, M.M. through ordinary and partial differential equations, M.M. through ordinary and partial difference equation, M.M. through integral equations, M.M. through integro-differential equations, M.M. through differential-difference equations, M.M. through functional equations, M.M. through graphs, M.M. through mathematical programming, M.M. through calculus of variations, M.M. through maximum principle and so on.

Again there are books on each of these techniques. However in most of these books, most of the space is devoted to explaining the theory of the technique concerned and applications are given as illustrations only. Mathematical modelling aspect is seldom emphasised.

In books of category (a), mathematical modelling is emphasised and techniques are considered of secondary importance (through this is not always the case) but the models belong to one specified field of knowledge. In books of category (b) the theory of the technique is emphasised and ready-made models are used to illustrate the technique. In the present book, we assume the knowledge of the basic theory of each technique and lay emphasis mainly on mathematical modelling and applications of the technique. In particular we consider when models in terms of specific techniques may be relevant. Books of category (a) consider applications of mathematics in one specified field of knowledge, but use a diversity of mathematical techniques. Books of category (b) use a single technique, but consider application in a diversity of fields of knowledge. In the present book, we consider both a diversity of techniques and a diversity of fields of knowledge.

(c) Mathematical Models may also be classified according to the *purpose* we have for the model. Thus we have Mathematical Models (M.M.) for Description, M.M. for Insight, M.M. for Prediction, M.M. for Optimization, M.M. for Control and M.M. for Action.

(d) Mathematical Models may also be classified according to their *nature*. Thus

(i) Mathematical Models may be *Linear* or *Non-Linear* according as the basic equations describing them are linear or non-linear.

(ii) Mathematical Models may be *Static or Dynamic* according as the time-variations in the system are not or are taken into account.

(iii) Mathematical Models may be *Deterministic or Stochastic* according as the chance factors are not or are taken into account.

(iv) Mathematical Models may be *Discrete or Continuous* according as the variables involved are discrete or continuous.

Linear, static and deterministic models are usually easier to handle than non-linear, dynamic and stochastic models and in general in any discipline, these are the first to be considered.

Continuous-variate models appear to be easier to handle than the discrete-variate models, due to the development of calculus and differential equations. In fact in many disciplines, these were developed first. However continuous models are simpler only when analytical solutions are available, otherwise we have to approximate a continuous, model also by a discrete model so that these can be handled numerically.

There are of course models which involve both discrete and continuous variates simultaneously.

Essentially most realistic models are non-linear, dynamic and stochastic. We use linear, static or deterministic models because these are easier to handle and give good approximate answers to our problems.

When the variables are essentially discrete, we may still use continuous models to be able to use calculus and differential equations. Similarly when the variables are essentially continuous, we may still use a discrete model to be able to use computers.

EXERCISE 1.3

1. Give two examples of each type of model mentioned in this section
2. Classify the following models

(i) $S = a + bp + cp^2$, $D = \alpha + \beta p + \gamma p^2$

[S is supply, D is demand, p is price]

(ii) $\dfrac{dx}{dt} = ax - bx^2$

[$x(t)$ is population at time t]

(iii) $x(t + 1) = ax(t) - bx(t)y(t)$

$y(t + 1) = -py(t) + qx(t)y(t)$

[$x(t), y(t)$ are population of prey and predator species respectively.]

(iv) $\dfrac{dp_n}{dt} = \lambda p_{n-1}(t) - \mu p_{n+1}(t) - (\lambda + \mu)p_n(t);$ $n = 1, 2, 3,$

[$p_n(t)$ is the probability of n persons at time t]

1.4 SOME CHARACTERISTICS OF MATHEMATICAL MODELS

(i) *Realism of models*: We want a mathematical model to be as realistic as possible and to represent reality as closely as possible. However, if a model is very realistic, it may not be mathematically tractable. In making a mathematical model, there has to be a trade-off between tractability and reality.

(ii) *Hierarchy of models*: Mathematical modelling is not a one-shot affair. Models are constantly improved to make them more realistic. Thus for every situation, we get a hierarchy of models, each more realistic than the preceding and each likely to be followed by a better one.

(iii) *Relative precision of models*:   Different models differ in their precision and their agreement with observations.

(iv) *Robustness of models*:   A mathematical model is said to be robust if small changes in the parameters lead to small changes in the behaviour of the model. The decision is made by using sensitivity analysis for the models.

(v) *Self-consistency of models*:   A mathematical model involves equations and inequations and these must be consistent, e.g. a model cannot have both $x + y > a$ and $x+y < a$. Sometimes the inconsistency results from inconsistency of basic assumptions. Since mathematical inconsistency is relatively easier to find out, this gives a method of finding inconsistency in requirements which social or biological scientists may require of their models. A well-known example of this is provided by Arrow's Impossibility Theorem.

(vi) *Oversimplified and overambitious models*:   It has been said that mathematics that is certain does not refer to reality and mathematics that refers to reality is not certain. A model may not represent reality because it is oversimplified. A model may also be overambitious in the sense that it may involve too many complications and may give results accurate to ten decimal places whereas the observations may be correct to two decimal places only.

(vii) *Complexity of models*:   This can be increased by subdividing variables, by taking more variables and by considering more details. Increase of complexity need not always lead to increase of insight as after a stage, diminishing returns begin to set in. The art of mathematical modelling consists in stopping before this stage.

(viii) *Models can lead to new experiments, new concepts and new mathematics*:   Comparison of predictions with observations reveals the need for new experiments to collect needed data. Mathematical models can also lead to development of new concepts. If known mathematical techniques are not adequate to deduce results from the mathematical model, new mathematical techniques have to be developed.

(iv) *A model may be good, adequate, similar to reality for one purpose and not for another*:   Thus we may need different models for explaining different aspects of the same situation or even for different ranges of the variables. Of course in this case, search for a unified model continues.

(x) *Models may lead to expected or unexpected predictions or even to nonsense*:   Usually models give predictions expected on common sense considerations, but the model predictions are more quantitative in nature. Sometimes they give unexpected predictions and then they may lead to break-throughs or deep thinking about assumptions. Sometimes models give prediction completely at variance with observations and then these models have to be drastically revised.

(xi) *A model is not good or bad; it does or does not fit*:   Models may lead to nice and elegant mathematical results, but only those models are acceptable which can explain, predict or control situations. A model may also fit one situation very well and may give a hopeless fit for another situation.

(xii) *Modelling forces us to think clearly*:   Before making a mathematical model, one has to be clear about the structure and essentials of the situation.

(xiii) *Sticking to one model may prevent insight*:   A model helps thinking, but it can also direct thinking in one narrow channel only. Sometimes insight is obtained by breaking with traditional models and designing entirely new ones with new concepts.

(xiv) *Inadequate models are also useful*:   Since they lead us to search for aspects which may have been neglected at first. Failures can be prelude to successes if we can find the reasons for these failures.

(xv) *Non-feedback models are improper*:   A model must include the possibility of its improvement in the light of the experimental or observational data.

(xvi) *Partial modelling for subsystems*:   Before making a model for the whole system, it may be convenient to make partial models for subsystems, test their validity and then integrate these partial models into a complete model. Sometimes existing models are combined to give models for bigger systems. Often models are unified so that the general model includes the earlier models as special cases.

(xvii) *Modelling in terms of modules*:   One may think of models for small modules and by combining them in different ways, one may get models for a large number of systems.

(xviii) *Imperfections of models and cost of modelling*:   No model is perfect and every model can be improved. However each such improvement may cost time and money. The improvement in the model must justify the investment made in this process.

(xix) *State variables and relations*:   For making a mathematical model, one has first to identify the state variables and then specify the relations between them. The right choice of state variables is of the utmost importance.

(xx) *Estimation of parameters*:   Every model contains some parameters and these have to be estimated. The model must itself suggest experiments or observations and the method of calculation of these parameters. Without this explicit specification, the model is incomplete.

(xxi) *Validation by independent data*: Sometimes parameters are estimated with the help of some data and the same data are used to validate the model. This is illegitimate. Independent data should be used to validate the model.

(xxii) *New models to simplify existing complicated models*: We start with simple models, introduce more and more variables and more and more functions to make the models more realistic and more complicated and with the additional insights obtained, we should again be able to simplify the complex models.

(xxiii) *Modelling $\Rightarrow$ Mathematics $+$ Discipline*: For making a mathematical models of a situation, one must know both mathematics and the discipline in which the situation arises. Efforts to make a mathematical model without deeply understanding the discipline concerned may lead to infructous models. Discipline insight must both precede and follow mathematical modelling.

(xxiv) *Transferability of mathematical models*: A mathematical model for one field may be equally valid for another field and may be validly transferred to another field, but great care must be exercised in this process. A model which is transferable to a number of fields is very useful, but no model should be thrust on a field unless it is really applicable there.

(xxv) *Prediction-validation-iteration cycle*: A mathematical model predicts conclusions which are then compared with observations. Usually there is some discrepancy. To remove this discrepancy, we improve the model, again predict and again try to validate and this iteration is repeated till a satisfactory model is obtained.

(xxvi) *Models for strategic and tactical thinking*: Models may be constructed for determining guidelines for particular situations or they may be for determining an overall strategy applicable to a variety of situations.

(xxvii) *Constraints of additivity and normality*: Models which are linear, additive and in which the probability distribution follows the normal law are relatively simpler, but relatively more realistic models have to be free from these constraints.

(xxviii) *Mathematical modellings and mathematical techniques*: Emphasis in applied mathematics has very often been on mathematical techniques, but the heart of applied mathematics is mathematical modelling.

(xxix) Mathematical modelling gives new ideology and unity *to applied mathematics*: Thus operations research and fluid dynamics differ in their subject matter as well as in techniques, but mathematical modelling is common to both.

(xxx) *Non-uniqueness of models*: A situation need not have only one mathematical model and the existence of one model for it should not inhibit search for better and different models.

(xxxi) *Dictionary of mathematical models*: It is unlikely that we shall ever have a complete dictionary of mathematical models so that our task will be only to choose an appropriate model for a given situation. Familiarity with existing models will always be useful, but new situations will always demand construction of new models.

(xxxii) *No prefabrication of models*: Some pure mathematicians believe that every consistent logical structure will one day model some physical situation. This is likely to be an exception rather than the rule. There will always be a very large number of mathematical structures without corresponding physical models and there will always be physical situations without good mathematical models. Search has to go on in both directions. Mathematics for modelling has to be mainly motivated by the world around us.

(xxxiii) *Mathematical modelling is an Art*: It requires experience, insight and understanding. Teaching this art is also another art.

(xxxiv) *Criteria for successful models*: These include good agreement between predictions and observations, of drawing further valid conclusions, simplicity of the model and its precision.

(xxxv) *Generality and applicability of models:* Laplace equation model applies to gravitational potential, electro-static potential, irrotational flows and a variety of other situations. There are some models applicable to a wide variety of situations, while there are others which are applicable to specific situations only.

(xxxvi) *Unity of disciplines through mathematical modelling:* When a number of different situations are represented by the same mathematical model, it reveals a certain identity of structures of these situations. It can lead to a certain economy of efforts and it can reveal a certain underlying unity between different disciplines.

EXERCISE 1.4

1. Illustrate each of the 36 characteristics given in this section with one or more mathematical models.

2. Take ten mathematical models given in this book and answer for each the following questions:

Is it linear or non-linear? Is it static or dynamic? Is it deterministic or stochastic? Is it for understanding or optimization or control? Does it apply to a number of fields? Is it realistic enough? How can it be made more realistic? Is it robust? Is it consistent? Is it oversimplified or overambitious? Does it suggest new ideas or new concepts? Which aspects of the situation does it explain and which it does not? Does it lead to expected or unexpected results? Can you solve it analytically or numerically? Can you give another model for the same situation? How would you estimate the parameters involved in the model? How would you validate the model? Can you simplify the model?

3. Elaborate each of the thirty-six concepts in ten or more lines.

4. Write a note one the significance of Arrow's Impossibility Theorem for Mathematically Modelling (cf. books 29,30 of Section 1.10).

5. Suggest some more questions of the type given in Ex. 2.

1:5 MATHEMATICAL MODELLING THROUGH GEOMETRY

(a) One of the earliest examples of mathematical modelling was that of mathematical description of the paths of planets. Looked at from the Earth the paths were not simple curves like circles or ellipses. The next curve known in order of complexity was an epicycloid which is the locus of a point on a circle which rolls on another fixed circle. The path of a planet was not even an epicycloid. However it was found possible to combine suitably a number of these epicycloidal curves or epicycles to describe the paths of all the planets. This was highly successful, though quite a complicated model.

(b) Another geometric modelling was involved in use of parabolic mirrors for burning enemy ships by Archimedes by concentrating Sun's

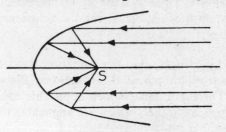

parallel rays on them. The property that was used is that the line joining a point P on a parabola to the focus S and the line though P parallel to the axis of the parabola make equal angles with the tangent (and the normal) at P so that all parallel rays of the Sun can be reflected to only one point, i.e. to the focus S (Figure 1.6).

Figure 1.6

(c) A similar geometric modelling is involved in constructing an elliptic sound gallery so that the sound produced at one focus can be heard at the other focus after being reflected back from every point of the ellipse (Figure 1.7).

(d) Based on the observations of Copernicus, Kepler showed that each

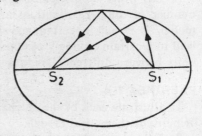

Figure 1.7

planet moves in an ellipse with the Sun at one focus. Thus the heliocentric theory of planetory motion completely simplified the description of the paths of the planets. The earlier geocentric theory required complicated combination of epicycloids. Both the models are correct, but the heliocentric model is much simpler than the geocentric model. However both the models were models for description only. Later Newton showed that the elliptical orbit followed from the universal law of gravitation and thus this model became a model for understanding. Still later in 1957, the elliptic orbits were used as orbits of satellites. At this stage, the model became a model for control. Now the same model can be used for getting optimal orbits for the satellites and as such it can also be used as a model for optimization.

(e) A fifth geometrical model is involved in the use of Fermat's principle of least time which states that light travels from one point to another in such a way as to take least possible time. One immediate consequence of this is that in a homogeneous medium, light travels in a straight line, since a straight line corresponds to the shortest distance between two points. If however light travels from point A to point B after being reflected from a mirror CD, the light ray will be incident at such a point O of the mirror (Figure 1.8) so that $AO + OB$ is minimum or such that $AO + OB'$ is minimum, where B' is the mirror-image of B in CD. But $AO + OB'$ is minimum when AOB' is a straight line. This gives

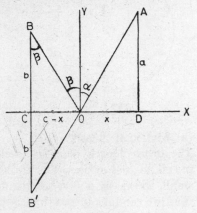

Figure 1.8

$$\angle AOY = \angle BOY \text{ or angle of incidence} = \text{angle of reflection} \qquad (3)$$

EXERCISE 1.5

1. Define a cycloid, epicycloid and hypocycloid and obtain their equations.

2. If a circle rolls over another which itself rolls over a third fixed circle, find the equation of the locus of a point of the first circle.

3. Prove the property of the parabola used in (b).

4. Prove the property of the ellipse used in (c).

5. Prove from Fermat's principle that if a light ray starts from a point within a parabola and is to be reflected back to the focus from a point on the mirror then, it must travel parallel to the axis of the parabola. Here a parabola is defined as the locus of a point whose distance from the focus is always equal to its distance from the directrise.

6. Find a curve such that a light ray starting from any point-inside it, after being reflected from any point on the curve, is always incident at a given point inside the curve.

1.6 MATHEMATICAL MODELLING THROUGH ALGEBRA

(a) Finding the Radius of the Earth

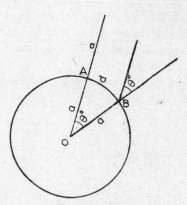

Figure 1.9

This model was used about two thousand years ago. A and B are two points on the surface of the Earth with the same longitude and d miles apart. When the Sun is vertically above A (i.e. it is in the direction OA, where O is the centre of the Earth) (Figure 1.9), the Sun's rays make an angle of $\theta°$ with the vertical at B (i.e. with the line OB). If a miles is the radius of the Earth, it is easily seen from Figure 1.9 that

$$\frac{d}{2\pi a} = \frac{\theta}{360} \quad \text{or} \quad a = \frac{360\, d}{2\pi\, \theta} \quad (4)$$

(b) Motion of Planets

The orbit of each planet is an ellipse with the Sun at one focus. However the ellipticities of the orbits are very small, so that as a first approximation, we can take these orbits as circles with Sun at the centre. Also we know that the planets move under gravitational attraction of the Sun and that for motion in a circle with uniform speed v, a central acceleration v^2/r is required.

If the masses of the Sun and the planet are S and P respectively, we get

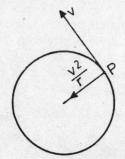

Figure 1.10

$$\frac{GPS}{r^2} = \frac{Pv^2}{r} \quad \text{or} \quad v^2 = \frac{GS}{r}, \quad (5)$$

where G is the constant of gravitation. Further if T is the periodic time of the planet, we have

$$vT = 2\pi r \quad (6)$$

Eliminating v between (5) and (6), we get

$$T^2 = \frac{4\pi^2 r^3}{GS} \quad (7)$$

If T_1, T_2 are the time periods of two planets with orbital radii r_1, r_2, then

$$T_1^2/T_2^2 = r_1^3/r_2^3, \tag{8}$$

so that the squares of the periodic times are proportional to the cubes of the radii of the orbits.

(c) Motions of Satellites

Satellites move under the attraction of the Earth in the same way as the planets move under the attraction of the Sun, so that we get

$$T^2 = \frac{4\pi^2 \, r^3}{GE}, \; \frac{T_1^2}{T_2^2} = \frac{r_1^3}{r_2^3} = \frac{(a + h_1)^3}{(a + h_2)^3} \tag{9}$$

where E is the mass of the Earth, a is the radius of the Earth and h_1, h_2 are the heights of the satellites above the Earth's surface. Also if g is the acceleration due to gravity at the Earth's surface, then

$$mg = \frac{G \, m \, E}{a^2} \quad \text{or} \quad GE = ga^2 \tag{10}$$

From (9) and (10)

$$T^2 = \frac{4\pi^2(a + h)^3}{ga^2}. \tag{11}$$

(d) We can solve the problem of 1.5 (e) by using algebraic method. In Figure 1.8,

$$m = AO + OB = (a^2 + x^2)^{1/2} + [b^2 + (c - x)^2]^{1/2} \tag{12}$$

Simplifying

$$4x^2 \left(1 - \frac{c^2}{m^2}\right) - 4cx \left(1 + \frac{k^2}{m^2}\right) + 4a^2 - \left(m + \frac{k^2}{m}\right)^2 = 0;$$

$$k^2 = a^2 - b^2 - c^2 \tag{13}$$

Since x is real, we get

$$c^2 \left(1 + \frac{k^2}{m^2}\right)^2 \geqslant \left(1 - \frac{c^2}{m^2}\right)\left(4a^2 - m^2\left(1 + \frac{k^2}{m^2}\right)^2\right)$$

or

$$[m^2 - (a + b)^2 - c^2]\,[m^2 - (a - b)^2 - c^2] \geqslant 0 \tag{14}$$

From Figure 1.8, the second factor is positive. As such (14) gives

$$m^2 \geqslant (a + b)^2 + c^2. \tag{15}$$

Thus the minimum value of m is $[(a + b)^2 + c^2]^{1/2}$ and when m has this value, the two roots of the quadratic (13) are equal and each is given by

$$2x = \frac{4c\left(1 + \dfrac{k^2}{m^2}\right)}{4\left(1 - \dfrac{c^2}{m^2}\right)} \quad \text{or} \quad \frac{2ac}{a + b} \tag{16}$$

so that

$$\frac{x}{a} = \frac{c-x}{b} \quad \text{or} \quad \alpha = \beta. \tag{17}$$

(e) By using the algebraic result that the arithmetic mean of n positive numbers $\geqslant$ the geometric mean of these numbers and the equality sign holds iff the numbers are equal, we can deduce that

(i) If the sum of n positive numbers is constant, then their product is maximum when the numbers are equal.

(ii) If the product of n positive numbers is constant, then their sum is minimum when the numbers are equal.

(iii) Of all rectangles with a given perimeter, the square has the maximum area.

(iv) Of all rectangles with a given area, the square has the minimum perimeter.

(v) Of all rectangular parallelopipeds with a given perimeter, the cube has the maximum volume.

(vi) Of all rectangular parallelopipeds with a given volume, the cube has the minimum perimeter.

(vii) The quantity $x^p(a-x)^q$ is maximum when $x/p = (a-x)/q$.

(f) In the same way, we have

$$w_1 x_1 + w_2 x_2 + \ldots + w_n x_n \geqslant x_1^{w_1} x_2^{w_2} \ldots x_n^{w_n}, \tag{18}$$

where $w_1, w_2, \ldots, w_n$ are positive weights with

$$\sum_{i=1}^{n} w_i = 1$$

and the equality sign holds iff $x_1 = x_2 = \ldots = x_n$

Now let $(p_1, p_2, \ldots, p_n)$, $(q_1, q_2, \ldots, q_n)$ be two probability distributions such that

$$\sum_{i=1}^{n} p_i = 1, \; \sum_{i=1}^{n} q_i = 1, p_i > 0, q_i > 0 \tag{19}$$

Putting $w_i = q_i$, $x_i = p_i/q_i$ in (18), we get

$$\sum_{i=1}^{n} q_i \frac{p_i}{q_i} \geqslant \sum_{i=1}^{n} \left(\frac{p_i}{q_i}\right)^{q_i}$$

or

$$\sum_{i=1}^{n} q_i \ln \frac{p_i}{q_i} \leqslant \ln \sum_{i=1}^{n} p_i = 0, \tag{20}$$

so that

$$\sum_{i=1}^{n} q_i \ln \frac{q_i}{p_i} \geqslant 0, \; \sum_{i=1}^{n} p_i \ln \frac{p_i}{q_i} \geqslant 0 \tag{21}$$

and the equality sign holds iff $p_i = q_i$ for all i. The inequality (21) is known as *Shannon's inequality* and is useful in the development of maximum-entropy models.

(g) Putting $q_i = \dfrac{1}{n}$ in (21), we get

$$\sum_{i=1}^{n} p_i \ln p_i + \sum_{i=1}^{n} p_i \ln n \geqslant 0 \quad \text{or} \quad \sum_{i=1}^{n} p_i \ln p_i \geqslant - \ln n$$

or

$$- \sum_{i=1}^{n} p_i \ln p_i \leqslant \ln n = - \sum_{i=1}^{n} \frac{1}{n} \ln \frac{1}{n} \tag{22}$$

The expression

$$S = - \sum_{i=1}^{n} p_i \ln p_i$$

is called the *entropy* of the probability distribution $(p_1, p_2, \ldots, p_n)$ and (22) shows that its maximum value is $\ln n$ and this maximum value is attained when all the probabilities are equal.

This result is of great importance in information theory, coding theory and in the development of maximum-entropy models.

EXERCISE 1.6

1. Given that for the Earth $T = 365$ days $= 3.15 \times 10^7$ secs,

$r = 93$ million miles $= 1.5 \times 10^{11}$ metres and

$G = 6.67 \times 10^{-11}$ nt $-$ m^2/kg^2,

use (7) to show that $S \simeq 2.0 \times 10^{30}$ kg.

2. Given the following data, about the solar system, use (7) to find the mass of the Sun from the data about each planet.

	Mean distance from Sun in 10^6 miles	Orbital period in years	Mass relative to Earth	Density relative to water	Diameter in miles
Sun	—	—	331100	1.39	865980
Mercury	36.0	0.24	0.34	4.80	2774
Venus	67.2	0.62	0.82	5.00	7566
Earth	93.0	1.00	1.00	5.53	7927
Mars	141.7	1.88	0.11	3.96	4216
Jupiter	483.5	11.86	314.5	1.34	87700
Saturn	886	29.46	94.1	0.70	76340
Uranus	1784	84.02	14.4	1.30	30880
Neptune	2794	164.79	16.7	1.40	32940
Pluto	3675	248.00	—	—	—
Moon	—	—	0.012	3.4	2162

3. Verify the formula (8) for the data of Ex. 2.

4. Determine the mass of the Earth from the data: Period T of mean orbit of Moon about the earth $= 27.3$ days and $r = 2.30 \times 10^5$ miles.

5. Find the value of g on the surface of each of the planets.

6. Use (ii) to find the height of the satellite above the Earth's surface if its time period is the same as that of the Earth viz. 24 hours.

7. Give the proof of the result (d) in complete detail.

8. Prove all the seven results stated in (e).

9. Prove that if $f(x)$ and $g(x)$ are probability density functions for a continuous variate over the range $[a, b]$, then

$$\int_a^b f(x) \ln \frac{f(x)}{g(x)} \, dx \geqslant 0$$

10. Prove that

$$-\int_a^b f(x) \ln f(x) \, dx$$

is maximum subject to

$$\int_a^b f(x) \, dx = 1 \quad \text{when} \quad f(x) = \frac{1}{b - a}.$$

1.7 MATHEMATICAL MODELLING THROUGH TRIGONOMETRY

(a) Finding the Distance of the Moon

From two points A, B on the surface of the Earth, will be same longitude, one in the Northern hemisphere and the other in the Southern hemisphere, measure angles θ_1, θ_2 between verticals at A and B and the directions of the centre of the Moon (Figure 1.11).

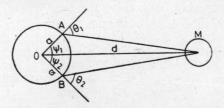

Figure 1.11

If d is the distance of the centre of the Moon's disc from the centre of Earth, Figure 1.11 gives

$$\frac{d}{\sin \theta_1} = \frac{a}{\sin (\theta_1 - \psi_1)}, \quad \frac{d}{\sin \theta_2} = \frac{a}{\sin (\theta_2 - \psi_2)} \tag{23}$$

Also

$$\psi_1 + \psi_2 = \alpha = \varphi_1 + \varphi_2 \tag{24}$$

where φ_1 is the Northern latitude of A and φ_2 is the Southern latitude of B. Since φ_1, φ_2 are known, $\psi_1 + \psi_2$ is known. Eliminating ψ_1, ψ_2 from (23) and (24), we get d in terms of a, θ_1, θ_2 which are all known.

(b) Finding the Distance of a Star

For a star, the base line provided by AB is too small and we choose the largest distance available to us viz. the positions of the Earth six months apart. We can measure angles SAX and SBX and since we know length AB which is about 186 million miles, we can determine AX and BX from traingle ABX (Figure 1.12).

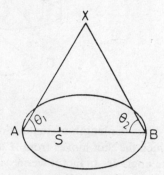

Figure 1.12

(c) Finding Length of the Day

The length of the day depends on the latitude of the place and declination of the Sun on the particular day.

In astronomy, we are concerned both with directions and distances of heavenly bodies. In spherical astronomy, we are concerned only with their directions from the observer.

We draw a sphere with unit radius with the observer at its centre. This sphere will be called the celestial sphere. If we join the observer to any heavenly body, this line will meet the celestial sphere on some point which will be called the position of the heavenly body on the celestial sphere.

If A, B, C are positions of three heavenly bodies on the celestial sphere and we join these by great circle arcs, we get a spherical triangle ABC, the lengths of whose sides are measured by the angles subtended by these arcs at the centre. The relations between sides and angles of a spherical triangles are given by

$$\frac{\sin A}{\sin a} = \frac{\sin B}{\sin b} = \frac{\sin C}{\sin c} \tag{25}$$

and

$$\cos a = \cos b \cos c + \sin b \sin c \cos A \tag{26}$$

The point vertically above the observer is called his zenith and is denoted by Z. Thus OZ gives the direction of the line joining the centre of the Earth to the observer. Similarly the direction of the axis of rotation of the Earth determines a point P on the celestial sphere and the arc PZ is equal to the angle between OP and OZ and is thus equal to $\pi/2 - \varphi$, where φ is the latitude of the observer.

Due to the daily rotation of the Earth about its axis OP, the Sun appears to move in a plane perpendicular to OP (Figure 1.13). The Sun rises above the horizon at A, reaches its highest position at B, sets at the point C and remains invisible during its motion on arc CDA.

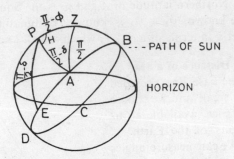

Figure 1.13

Thus the length of the day corresponds to the length of the time during which the Sun moves from A to B and from B to C and is determined by the angle H. In fact if we measure H in radians, then the change in H by 2π radians corresponds to a duration of 24 hours.

In spherical triangle ZPA, $ZP = \pi/2 - \varphi$, $ZA = \pi/2$ (since Z is the pole of the horizon and is distant $\pi/2$ from every point on the horizon). Also $PA = \pi/2 - \delta$, where δ is the declination of the Sun. Using (26) in spherical triangle ZPA, we get

$$\cos \frac{\pi}{2} = \cos \left(\frac{\pi}{2} - \varphi\right) \cos\left(\frac{\pi}{2} - \delta\right)$$

$$+ \sin \left(\frac{\pi}{2} - \varphi\right) \sin \left(\frac{\pi}{2} - \delta\right) \cos H$$

or

$$\cos H = -\tan \varphi \tan \delta \tag{27}$$

If φ and δ are known, then (27) determines H and then the length of the day $= 2H \times \dfrac{24}{2\pi} = 24 \dfrac{H}{\pi}$ hours.

The declination of the Sun varies from $-23\frac{1}{2}^{\circ} = -\epsilon$ to $23\frac{1}{2}^{\circ} = \epsilon$. In fact

$$\delta = -23\frac{1}{2}^{\circ} \text{ on 23rd December,} \qquad \delta = 0 \text{ on 21st March,}$$

$$\delta = 23\frac{1}{2}^{\circ} \text{ on 23rd June,} \qquad \delta = 0 \text{ on 21st September.}$$

In the morning, the twilight starts when the Sun is at E where $ZE = 108^{\circ}$ and $\angle ZPE = H'$ so that we get from the angle ZPE, using (26),

$$\cos 108^{\circ} = \sin \varphi \sin \delta + \cos \varphi \cos \delta \cos H' \tag{28}$$

From (27) and (28), we can find H and H' and therefore $H' - H$. The total length of the two twilights in the morning and evening is $24(H' - H)\pi$ hours.

EXERCISE 1.7

1. Show that the Sun does not set at a place if $\varphi > \dfrac{\pi}{2} - \delta$.

2. Show that the Sun does set at a place if $\varphi < \dfrac{\pi}{2} - \epsilon$.

3. Show that at places between latitudes $-\left(\dfrac{\pi}{2} - \epsilon\right)$ and $\left(\dfrac{\pi}{2} - \epsilon\right)$, the length of the day on 21st March and 21st September is 12 hours.

4. Show that at all these places the shortest day occurs on 21st December and the longest day occurs on 21st June.

5. Find the shortest and longest days at latitudes $0°$, $10°$, $20°$, $30°$, $40°$, $50°$, $60°$, $70°$, $80°$.

6. Assuming that δ changes uniformly throughout the year, find the length of the day in your town on the 21st day of every month.

7. If evening twilight ends when the Sun's centre is $18°$ below the horizon, show that at the equator, the duration of evening twilight is given in hours by

$$\frac{12}{\pi} \sin^{-1}(\sin 18° \sec \delta)$$

Use this formula to calculate the duration of evening twilight at the summer solstice.

8. Show that at a place is in latitude φ, the shortest duration of twilight expressed in hours is

$$\frac{2}{15} \sin^{-1}(\sin 9° \sec \varphi)$$

where $\sin^{-1}(\sin 9° \sec \varphi)$ is expressed in degrees.

9. If twilight begins or ends when the Sun is $18°$ below the horizon, show that all places have a day of more than 12 hours, including twilight, so long as the declination of the Sun is less than $18°$.

10. Explain how you would find the diameters of all the planets.

1.8 MATHEMATICAL MODELLING THROUGH CALCULUS

(a) Law of Reflection

Consider the model of Section 1.5(e) and 1.6(d). The distance travelled by light from A to B is

$$m = \sqrt{a^2 + x^2} + \sqrt{b^2 + (c - x)^2}, \tag{29}$$

so that

$$\frac{dm}{dx} = \frac{x}{\sqrt{a^2 + x^2}} - \frac{c - x}{\sqrt{b^2 + (c - x)^2}} \tag{30}$$

$$\frac{d^2m}{dx^2} = \frac{a^2}{(a^2 - x^2)^{3/2}} + \frac{b^2}{(b^2 + (c - x)^2)^{3/2}} > 0, \tag{31}$$

so that m is minimum when, using Figure 1.8,

$$\frac{x}{\sqrt{a^2 + x^2}} = \frac{c - x}{\sqrt{b^2 + (c - x)^2}} \quad \text{or} \quad \sin \alpha = \sin \beta \tag{32}$$

(b) Law of Refraction of Light

Consider the problem of refraction of light from a point A in vacuum to a point B (Figure 1.14) in a medium of refractive index μ. If light travels with velocity V in vacuum, it travels with velocity V/μ in the second medium so that the time T of travel is given by

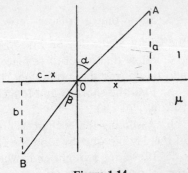

Figure 1.14

$$T = \frac{\sqrt{a^2 + x^2}}{V} + \mu \frac{\sqrt{b^2 + (c - x)^2}}{V} \tag{33}$$

so that

$$V\frac{dT}{dx} = \frac{x}{\sqrt{a^2 + x^2}} - \mu \frac{c - x}{\sqrt{b^2 + (c - x)^2}} \tag{34}$$

$$V\frac{d^2T}{dx^2} = \frac{a^2}{(a^2 + x^2)^{3/2}} + \frac{\mu b^2}{(b^2 + (c - x)^2)^{3/2}} \tag{35}$$

Thus T is minimum when

$$\frac{x}{\sqrt{a^2 + x^2}} = \mu \frac{c - x}{\sqrt{b^2 + (c - x)^2}} \quad \text{or} \quad \sin \alpha = \mu \sin \beta \tag{36}$$

(c) EOQ Model for Inventory Control

Let the total demand for a commodity be D units in a year and let orders of q units be placed D/q times in a year. Let the cost of ordering be C_1 per order so that the total cost of ordering is $C_1 D/q$.

In each order interval of duration q/D, the quantity in stock falls uniformly from q to 0 so that the average stock in this period is $q/2$. Let the cost of storing be C_2 per unit item per unit time so that the total storage cost is $C_2 \frac{q}{2} \cdot \frac{q}{D} \frac{D}{q} = C_2 \frac{q}{2}$, so that the total cost of ordering and storing

is

$$f(q) = C_1 \frac{D}{q} + \frac{C_2}{2} q, \tag{37}$$

so that

$$f'(q) = - C_1 \frac{D}{q^2} + \frac{C_2}{2}, \quad f''(q) = \frac{2C_1 D}{q^3} > 0, \tag{38}$$

so that the total cost of storing and ordering is minimum when

$$q = \sqrt{\frac{2C_1 D}{C_2}}, \quad \frac{D}{q} = \sqrt{\frac{DC_2}{2C_1}} \tag{39}$$

Thus for minimizing the inventory cost, the quantity to be ordered each time is directly proportional to the square root of D and the square root of C_1 and is inversely proportional to the square root of C_2.

(d) Triangle of Given Perimeter with Maximum Area

The square of the area of a triangle with semi-perimeters is given by

$$\Delta^2 = s(s - a)(s - b)(s - c) = s^4 - s^3(a + b + c)$$
$$+ s^2(ab + bc + ca) - sabc$$

$$= s^4 - 2s^4 + \frac{s^2}{2}[4s^2 - a^2 - b^2 - c^2] - sabc$$

$$= s^4 - \frac{s^2}{2}(a^2 + b^2 + c^2) - sabc \tag{40}$$

This has to be maximized subject to

$$2s = a + b + c \tag{41}$$

Using Lagrange's method, this gives

$$\frac{s^2a + sbc}{1} = \frac{s^2b + sca}{1} = \frac{s^2c + sab}{1}$$

or

$$\frac{sa^2 + abc}{a} = \frac{sb^2 + abc}{b} = \frac{sc^2 + abc}{c}$$

$$= \frac{s(a^2 - b^2)}{a - b} = \frac{s(b^2 - c^2)}{b - c}$$

or

$$a + b = b + c \quad \text{or} \quad a = c \tag{42}$$

Similarly $\qquad b = c$ so that $a = b = c$ $\qquad\qquad$ (43)

Thus the triangle of maximum area is the equilateral triangle. Of course, the triangle of minimum area has zero area.

(e) Parallelopiped with Given Perimeter and Maximum Volume

We have to maximize xyz subject to $x + y + z = 3a$

The Lagrangian is

$$L = xyz - \lambda(x + y + z - 3a) \tag{44}$$

This is maximum when

$$yz = zx = xy \quad \text{or} \quad x = y = z \tag{45}$$

Thus the parallelopiped of maximum volume is a cube. The parallelopiped of minimum volume has of course zero volume.

Alternatively

$$V = xy(3a - x - y)$$
$$\frac{\partial V}{\partial x} = 3ay - 2xy - y^2, \frac{\partial V}{\partial y} = 3ax - x^2 - 2xy \tag{46}$$

$$\frac{\partial^2 V}{\partial x^2} = -2y, \; \frac{\partial^2 V}{\partial y^2} = -2x, \; \frac{\partial^2 V}{\partial x \partial y} = 3a - 2x - 2y \qquad (47)$$

Putting the first derivatives equal to zero, we get $x = y = a$ and at this point

$$\frac{\partial^2 V}{\partial x^2} = -2a < 0, \; \frac{\partial^2 V}{\partial y^2} = -2a < 0,$$

$$\frac{\partial^2 V}{\partial x^2} \frac{\partial^2 V}{\partial y^2} - \left(\frac{\partial^2 V}{\partial x \partial y}\right)^2 = a^2 > 0, \qquad (48)$$

so that the volume is maximum when $x = y = z = a$, i.e. when the rectangular parallelopiped is a cube.

(f) Mathematics of Business

(i) Let the revenue obtained by selling a commodity at price p be given by

$$R(p) = ap - bp^2, \qquad (49)$$

so that

$$R'(p) = a - 2bp, \quad R''(p) = -2b \qquad (50)$$

Thus the profit is maximum when the price is $a/2b$.

(ii) The revenue on selling x items in $R(x)$ and the cost of x items is $C(x)$, then the profit function $P(x)$ is given by

$$P(x) = R(x) - C(x), \qquad (51)$$

so that

$$P'(x) = R'(x) - C'(x), \quad P''(x) = R''(x) - C''(x) \qquad (52)$$

Thus the profit is maximum when marginal revenue (revenue from selling one additional item) is equal to marginal cost (cost of producing one additional item).

(iii) If the total cost of producing q item is $a + bq + cq^2$, then the average cost per item is

$$\varphi(q) = \frac{a}{q} + b + cq, \qquad (53)$$

so that

$$\varphi'(q) = -\frac{a}{q^2} + c, \quad \varphi''(q) = \frac{2a}{q^3}, \qquad (54)$$

and the average cost per item is minimum when $q = \sqrt{\dfrac{a}{c}}$.

(iv) If the profit in a district for x sales representatives is

$$f(x) = -ax^2 + bx - c \qquad (55)$$

then

$$f'(x) = -2ax + b, \quad f''(x) = -2a \qquad (56)$$

Thus the profit is maximum for $b/2a$ sales representations.

(v) If the number of travelling passengers in a city bus system is $a - bp$, where p is the price of a ticket, then the total revenue is $ap - bp^2$ and this is maximum when the price is $a/2b$.

(vi) The demands for two related products sold by a company are given by

$$q_1 = a_1 - a_2 p_1 - a_3 p_2 \quad \text{and} \quad q_2 = b_1 - b_2 p_1 - b_3 p_2 \quad (57)$$

and the total sales value is

$$S = p_1 q_1 + p_2 q_2 = a_1 p_1 + b_1 p_2 - a_2 p_1^2 - b_3 p_2^2$$
$$- a_3 p_1 p_2 - b_2 p_1 p_2 \quad (58)$$

For maximizing the sale value, we charge prices given by

$$a_1 - 2a_2 p_1 - (a_3 + b_2) p_2 = 0;$$
$$b_1 - 2b_3 p_2 - (a_3 + b_2) p_1 = 0 \quad (59)$$

EXERCISE 1.8

1. Two mirrors are placed along OX ane OY. A light ray starts from (a, b) and after being reflected from the two mirrors reaches the point (c, d). Prove that for the total time of travel to be minimum, angle of incidence is equal to angle of reflection at each mirror.

2. A light ray starts from a point A in one medium with refractive index μ_1 and after passing through a medium with refractive index μ_2 reaches a point B in a third medium with refractive index μ_3 (Figure 1.15). Show that

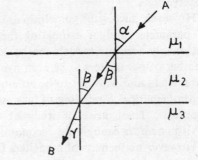

Figure 1.15

$$\mu_1 \sin \alpha = \mu_2 \sin \beta = \mu_3 \sin \gamma$$

3. In inventory control problem (c), we allow shortages to occur which can be supplied when items are received. If C_3 is the cost of shortage per unit per unit time, and S is the maximum shortage allowed, show that the inventory cost is

$$f(q, S) = \frac{D}{q} C_1 + \frac{(q - S)^2}{2q} C_2 + \frac{S^2 C_3}{2q}$$

Find values of q and S for minimizing this inventory cost.

4. Find the cyclic quadrilateral with given perimeters and maximum area (The area of a cyclic quadrilateral as proved first by Brahmgupta is

$$[S(S - a)(S - b)(S - c)(S - d)]^{1/2})$$

Can you extend this result to any quadrilateral or pentagon? Discuss. Also solve this exercise and problem of section 1.8(d) by using results of section 1.6(e).

5. Find the rectangular parallelopiped with given area of the faces and maximum volume.

6. A company wants to buy a rectangular plot of 10,000 square metres. The price is Rs. 50 per metre of length on the main road and Rs. 20 per metre on the other three sides. Find the sides of the cheapest plot the company can buy.

1.9 LIMITATIONS OF MATHEMATICAL MODELLING

There are thousands of mathematical models which have been successfully developed and applied to get insight into tens of thousands of situations. In fact mathematical physics, mathematical economics, operations research, biomathematics etc. are almost synonymous with mathematical modelling.

However there are still an equally large or even a larger number of situations which have not yet been mathematically modelled either because the situations are sufficiently complex or because mathematical models formed are mathematically intractable.

The development of powerful computers has enabled a much larger number of situations to be mathematically modelled. Moreover it has been possible to make more realistic models and to obtain better agreement with observations.

However, successful guidelines are not available for choosing the number of parameter and of estimating the values for these parameters. In fact reasonably accurate models can be developed to fit any data by choosing number of parameters to be even five or six. We want a minimal number of parameters and we want to be able to estimate them accurately.

Mathematical Modelling of large-scale systems presents its own special problems. These arise in study of world models and in global models of environment, oceanography, economic conditions, pollution control etc.

However mathematical modellers from all disciplines-mathematics, statistics, computer science, physics, engineering, social sciences—are meeting the challenges with courage. Six international conferences on Mathematical Modelling have been held and a large number of specialised conferences on mathematical modelling have been organised. Teaching of Mathematical Modelling has not been neglected and the first three international conference on the Teaching of Mathematical Modelling have already been held.

BIBLIOGRAPHY

1. J.G. Andrews and R.R. Mclone (1976). *Mathematical Modelling*, Butterworths, London.
2. R. Aris (1978). *Mathematical Modelling Techniques*, Pitman.
3. R. Aris and M. Penn (1980) The Mere Notion of a Model: Mathematical Modelling 1, 1-12.
4. X.J.R. Avula and E.Y. Roden (Editors) (1980). Mathematical Modelling: An International Journal, Pergamon Press, New York.
5. C.A. Bender (1978). An Introduction to Mathematical Modelling, Wiley Interscience, New York.
6. C.A. Brabbia and J.J. Connon (Editors) (1977). Applied Mathematical Modelling (Journal) IPC Science and Technology Press, Guildford.
7. D.N. Burghes and G.M. Read (Editors) (1977). Journal of Mathematical Modelling for teachers, Cranefield Institute of Technology.

8. H. Burkhardt (1979). Learning to Use Mathematics, Bull. Inst. Maths. App. **15**, 238-243.
9. H. Burkhardt (Editor) (1978). *Teaching Methods for Undergraduate Mathematics*, Shell Centre for Mathematics Education, Nottingham.
10. H. Burkhardt (1976). *The Real World and Mathematics*, Shell Centre for Mathematics Education, London.
11. H. Hahamann (1977). *Mathematical Models*, Prentice-Hall.
12. J.N. Kapur (1976). Mathematical Models in the Social Sciences. Bull. Math. Ass. Ind. **4**, 30-37.
13. J.N. Kapur (1976). Some Modern Applications of Mathematics. Bull. Math. Ass. Ind. **8**, 16-19.
14. J.N. Kapur (1978). Some Problems in Biomathematics. Int. Jour. Maths. Edu. Sci. Tech. **9**(3) 287-306.
15. J.N. Kapur (1979a). Mathematical Modelling, a New Identity for Applied Mathematics, Bull. Math. Ass. Ind. **11**, 45-47.
16. J.N. Kapur (1979b). Mathematical Modelling, its philosophy, scope, power and limitations, Bull, Math. Ass. Ind. **11**(3, 4) 62-112.
17. J.N. Kapur (1981). Some Aspects of Mathematical Modelling of Large-Scale Systems. Bull. Math. Ass, Ind. **11**, 24-43.
18. J.N. Kapur (1980a). The Art of Teaching the Art of Mathematical Modelling, Int. Journ. Math. Edu. Sci. Tech. **13**(2) 175-192.
19. J.N. Kapur (1980b). Some Mathematical Models in Population Dynamics, Kanpur Univ. Journ. Res. (Science) **1**, 1-161.
20. J.N. Kapur (1983). Twenty-five Years of Maximum-Entropy, Journ. of Math. and Phys. Science **17**(2) 103-156.
21. J.N. Kapur (1985). *Mathematical Models in Biology and Medicine*, Affiliated East-West Press, New Delhi.
22. J.N. Kapur (1988). *Maximum-Entropy Models in Science and Engineering*, New Delhi.
23. P. Lancaster (1976). *Mathematical Models of the Real World*, Prentice-Hall.
24. M.J. Lighthill (Editor) (1978). *Newer Uses of Mathematics*, Pergamon.
25. C.C. Lin and L.A. Segal (1974). *Mathematics Applied to Deterministic Problems in the Natural Sciences*, McMillan, New York.
26. D.I. Maki and Maynard Thompson (1973). *Mathematical Models and Applications*, Prentice-Hall.
27. R.R. Mclone (1979). Teaching Mathematical Modelling. Bull. Inst. Maths. and App. **15**, 244-246.
28. Ben Noble (1970). *Applications of Undergraduate Mathematics*, McMillan, New York.
29. R. Olinik (1978). *An Introduction to Mathematical Models in Social and Life Sciences*, Addison Wesley, New York.
30. F. Roberts (1976). *Discrete Mathematical Models*, Prentice-Hall.
31. J. Spanner (1980). Thoughts About the Essentials of Mathematical Modelling, Math. Mod. **1**, 93-108.

2

Mathematical Modelling Through Ordinary Differential Equations of First Order

2.1 MATHEMATICAL MODELLING THROUGH DIFFERENTIAL EQUATIONS

Mathematical Modelling in terms of differential equations arises when the situation modelled involves some *continuous* variable(s) varying with respect to some other continuous variable(s) and we have some reasonable hypotheses about the *rates of change* of dependent variable(s) with respect to independent variable(s).

When we have one dependent variable x (say population size) depending on one independent variable (say time t), we get a mathematical model in terms of an *ordinary differential equation* of the *first order*, if the hypothesis is about the rate of change dx/dt. The model will be in terms of an *ordinary differential equation of the second order* if the hypothesis involves the rate of change of dx/dt.

If there are a number of dependent continuous variables and only one independent variable, the hypothesis may give a mathematical model in terms of a *system of first or higher order ordinary differential equations*.

If there is one dependent continuous variable (say velocity of fluid u) and a number of independent continuous variables (say space coordinates x, y, z and time t), we get a mathematical model in terms of a *partial differential equation*. If there are a number of dependent continuous variables and a number of independent continuous variables, we can get a mathematical model in terms of systems of *partial differential equations*.

Mathematical models in terms of ordinary differential equations will be studied in this and the next two chapters. Mathematical models in terms of partial differential equations will be studied in Chapter 7.

2.2 LINEAR GROWTH AND DECAY MODELS

2.2.1 Populational Growth Models

Let $x(t)$ be the population size at time t and let b and d be the birth and death rates, i.e. the number of individuals born or dying per individual per unit time, then in time interval $(t, t + \Delta t)$, the numbers of births and deaths would be $bx \Delta t + 0(\Delta t)$ and $dx \Delta t + 0(\Delta t)$ where $0(\Delta t)$ is an infinitesimal which approaches zero as Δt approaches zero, so that

$$x(t + \Delta t) - x(t) = (bx(t) - dx(t))\Delta t + 0(\Delta t), \tag{1}$$

so that dividing by Δt and proceeding to the limit as $\Delta t \to 0$, we get

$$\frac{dx}{dt} = (b - d)x = ax \quad \text{(say)} \tag{2}$$

Integrating (2), we get

$$x(t) = x(0) \exp (at), \tag{3}$$

so that the population grows exponentially if $a > 0$, decays exponentially if $a < 0$ and remains constant if $a = 0$ (Figure 2.1)

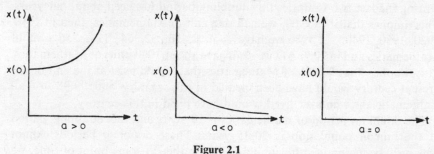

Figure 2.1

(i) If $a > 0$, the population will become double its present size at time T, where

$$2x(0) = x(0) \exp (aT) \quad \text{or} \quad \exp (aT) = 2$$

or
$$T = \frac{1}{a} \ln 2 = (0.69314118)a^{-1} \tag{4}$$

T is called the doubling period of the population and it may be noted that this doubling period is independent of $x(0)$. It depends only on a and is such that greater the value of a (i.e. greater the difference between birth and death rates), the smaller is the doubling period.

(ii) If $a < 0$, the population will become half its present size in time T' when

$$\frac{1}{2}x(0) = x(0) \exp (aT') \quad \text{or} \quad \exp (aT') = \frac{1}{2}$$

or
$$T' = \frac{1}{a} \ln \frac{1}{2} = -(0.69314118) a^{-1} \tag{5}$$

It may be noted that T' is also independent of $x(0)$ and since $a < 0$, $T' > 0$. T' may be called the half-life (period) of the population and it decreases as the excess of death rate over birth rate increases.

2.2.2 Growth of Science and Scientists

Let $S(t)$ denote the number of scientists at time t, $bS(t)\Delta t + 0(\Delta t)$ be the number of new scientists trained in time interval $(t \ t + \Delta t)$ and let $dS(t)\Delta t + 0(\Delta t)$ be the number of scientists who retire from science in the same period, then the above model applies and the number of scientists should grow exponentially.

The same model applies to the growth of Science, Mathematics and Technology. Thus if $M(t)$ is the amount of Mathematics at time t, then the rate of growth of Mathematics is proportional to the amount of Mathematics, so that

$$dM/dt = aM \qquad \text{or} \qquad M(t) = M(0) \exp(at) \qquad (6)$$

Thus according to this model, Mathematics, Science and Technology grow at an exponential rate and double themselves in a certain period of time. During the last two centuries this doubling period has been about ten years. This implies that if in 1900, we had one unit of Mathematics, then in 1910, 1920, 1930, 1940, ... 1980 we have 2, 4, 8, 16, 32, 64, 128, 256 unit of Mathematics and in 2000 AD we shall have about 1000 units of Mathematics. This implies that 99.9% of Mathematics that would exist at the end of the present century would have been created in this century and 99.9% of all mathematicians who ever lived, would have lived in this century.

The doubling period of mathematics is 10 years and the doubling period of the human population is 30-35 years. These doubling periods cannot obviously be maintained indefinitely because then at some point of time, we shall have more mathematicians than human beings. Ultimately the doubling period of both will be the same, but hopefully this is a long way away.

This model also shows that the doubling period can be shortened by having more intensive training programmes for mathematicians and scientists and by creating conditions in which they continue to do creative work for longer durations in life.

2.2.3 Effects of Immigration and Emigration on Population Size

If there is immigration into the population from outside at a rate proportional to the population size, the effect is equivalent to increasing the birth rate. Similarly if there is emigration from the population at a rate proportional to the population size, the effect is the same as that of increase in the death rate.

If however immigration and emigration take place at constant rate i and e respectively, equation (3) is modified to

$$\frac{dx}{dt} = bx - dx + i - e = ax + k \qquad (7)$$

Integrating (7) we get

$$x(t) + \frac{k}{a} = \left(x(0) + \frac{k}{a}\right)e^{at} \qquad (8)$$

The model also applies to growth of populations of bacteria and micro-organisms, to the increase of volume of timber in forest, to the growth of malignant cells etc. In the case of forests, planting of new plants will correspond to immigration and cutting of trees will correspond to emigration.

2.2.4 Interest Compounded Continuously

Let the amount at time t be $x(t)$ and let interest at rate r per unit amount per unit time be compounded continuously then

$$x(t + \Delta t) = x(t) + rx(t)\Delta t + 0(\Delta t),$$

giving

$$\frac{dx}{dt} = xr; \quad x(t) = x(0)e^{rt} \tag{9}$$

This formula can also be derived from the formula for compound interest

$$x(t) = x(0)\left(1 + \frac{r}{n}\right)^{nt}, \tag{10}$$

when interest is payable n times per unit time, by taking the limit as $n \to \infty$. In fact comparison of (9) and (10) gives us two definitions of the trancendental number e viz.

(i) e is the amount of an initial capital of one unit invested for one unit of time when the interest at unit rate is compounded continuously

(ii) $\quad e = \underset{n \to \infty}{\mathrm{Lt}} \left(1 + \frac{1}{n}\right)^{n}$ \hfill (11)

Also from (9) if $x(t) = 1$, then

$$x(0) = e^{-rt}, \tag{12}$$

so that e^{-rt} is the present value of a unit amount due one period hence when interest at the rate r per unit amount per unit time is compounded continuously.

2.2.5 Radio-Active Decay

Many substances undergo radio-active decay at a rate proportional to the amount of the radioactive substance present at any time and each of them has a half-life period. For uranium 238 it is 4.55 billion years. For potassium it is 1.3 billion years. For thorium it is 13.9 billion years. For rubidium it is 50 billion years while for carbon 14, it is only 5568 years and for white lead it is only 22 years.

In radiogeology, these results are used for radioactive dating. Thus the ratio of radio-carbon to ordinary carbon (carbon 12) in dead plants and animals enables us to estimate their time of death. Radioactive dating has also been used to estimate the age of the solar system and of earth as 45 billion years.

2.2.6 Decrease of Temperature

According to Newton's law of cooling, the rate of change of temperature of a body is proportional to the difference between the temperature T of the body and temperature T_s of the surrounding medium, so that

$$\frac{dT}{dt} = k(T - T_s), \quad k < 0 \tag{13}$$

and

$$T(t) - T_s = (T(0) - T_s)e^{kt} \tag{14}$$

and the excess of the temperature of the body over that of the surrounding medium decays exponentially.

2.2.7 Diffusion

According to Fick's law of diffusion, the time rate of movement of a solute across a thin membrane is proportional of the area of the membrane and to the difference in concentrations of the solute on the two sides of the membrane.

If the area of the membrane is constant and the concentration of solute on one side is kept fixed at a and the concentration of the solution on the other side initially is $c_0 < a$, then Fick's law gives

$$\frac{dc}{dt} = k(a - c), \quad c(0) = c_0, \tag{15}$$

so that

$$a - c(t) = (a - c(0))e^{-kt} \tag{16}$$

and $c(t) \to a$ as $t \to \infty$, whatever be the value of c_0.

2.2.8 Change of Price of a Commodity

Let $p(t)$ be the price of a commodity at time t, then its rate of change is proportional to the difference between the demand $d(t)$ and the supply $s(t)$ of the commodity in the market so that

$$\frac{dp}{dt} = k(d(t) - s(t)), \tag{17}$$

where $k > 0$, since if demand is more than the supply, the price increases. If $d(t)$ and $s(t)$ are assumed linear functions of $p(t)$, i.e. if

$$d(t) = d_1 + d_2 p(t), \quad s(t) = s_1 + s_2 p(t), \quad d_2 < 0, s_2 > 0 \tag{18}$$

we get

$$\frac{dp}{dt} = k(d_1 - s_1 + (d_2 - s_2)p(t)) = k(a - \beta p(t)), \quad \beta > 0 \tag{19}$$

or

$$\frac{dp}{dt} = K(p_e - p(t), \tag{20}$$

where p_e is the equilibrium price, so that

$$p_e - p(t) = (p_e - p(0))e^{-kt} \qquad (21)$$

and

$$p(t) \to p_e \quad \text{as} \quad t \to \infty$$

EXERCISE 2.2

1. Suppose the population of the world now is 4 billion and its doubling period is 35 years, what will be the population of the world after 350 years, 700 years, 1050 years? If the surface area of the earth is 1,860,000 billion square feet, how much space would each person get after 1050 years?

2. Find the relation between doubling, tripling and quadrupling times for a population.

3. In an archeological wooden specimen, only 25% of original radio carbon 12 is present. When was it made?

4. The rate of change of atmospheric pressure p with respect to height h is assumed proportional to p. If $p = 14.7$ psi at $h = 0$ and $p = 7.35$ at $h = 17,500$ feet, what is p at $h = 10,000$ feet?

5. What is the rate of interest compounded continuously if a bank's rate of interest is 10% per annum?

6. A body where temperature T is initially 300°C is placed in a large block of ice. Find its temperature at the end of 2 and 3 minutes?

7. The concentration of potassium in kidney is 0.0025 milligrammes per cubic centimetre. The kidney is placed in a large vessel in which the potassium concentration is 0.0040 mg/cm^3. In 1 hour the concentration in the kidney increases to 0.0027 mg/cm^3. After how much time will the concentration be 0.0035 mg/cm^3?

8. A population is decaying exponentially. Can this decay be stopped or reversed by an immigration at a large constant rate into the population?

2.3 NON-LINEAR GROWTH AND DECAY MODELS

2.3.1 Logistic Law of Population Growth

As population increases, due to overcrowding and limitations of resources, the birth rate b decreases and the death rate d increases with the population size x. The simplest assumption is to take

$$b = b_1 - b_2 x, \, d = d_1 + d_2 x, \quad b_1, b_2, d_1, d_2 > 0, \qquad (22)$$

so that (2) becomes

$$\frac{dx}{dt} = ((b_1 - d_1) - (b_2 + d_2)x) = x(a - bx), \, a > 0, b > 0 \qquad (23)$$

Integrating (23), we get

$$\frac{x(t)}{a - bx(t)} = \frac{x(0)}{a - bx(0)} e^{at} \qquad (24)$$

Equations (23) and (24) show that

(i) $x(0) < a/b \Rightarrow x(t) < a/b \Rightarrow dx/dt > 0 \Rightarrow x(t)$ is a monotonic increasing function of t which approaches a/b as $t \to \infty$.

(ii) $x(0) > a/b \Rightarrow x(t) > a/b \Rightarrow dx/dt < 0 \Rightarrow x(t)$ is a monotonic decreasing function of t which approaches a/b as $t \to \infty$.

Now from (23)

$$\frac{d^2x}{dt^2} = a - 2bx,\qquad(25)$$

so that $d^2x/dt^2 \gtreqless 0$ according as $x \lesseqgtr a/2b$. Thus in case (i) the growth curve is convex if $x < a/2b$ and is concave if $x > a/2b$ and it has a point of inflexion at $x = a/2b$. Thus the graph of $x(t)$ against t is as given in Figure 2.2.

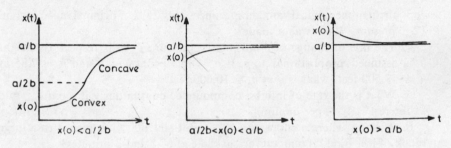

Figure 2.2

—If $x(0) < a/2b$, $x(t)$ increases at an increasing rate till $x(t)$ reaches $a/2b$ and then it increases at a decreasing rate and approaches a/b at $t \to \infty$

—If $a/2b < x(0) < a/b$, $x(t)$ increases at a decreasing rate and approaches a/b as $t \to \infty$

—If $x(0) = a/b$, $x(t)$ is always equal to a/b

—If $x(0) > a/b$, $x(t)$ decreases at a decreasing absolute rate and approaches a/b as $t \to \infty$

2.3.2 Spread of Technological Innovations and Infectious Diseases

Let $N(t)$ be the number of companies which have adopted a technological innovation till time t, then the rate of change of the number of these companies depends both on the number of companies which have adopted this innovation and on the number of those which have not yet adopted it, so that if R is the total number of companies in the region

$$\frac{dN}{dt} = kN(R - N),\qquad(26)$$

which is the logistic law and shows that ultimately all companies will adopt this innovation.

Similarly if $N(t)$ is the number of infected persons, the rate at which the number of infected persons increases depends on the product of the numbers of infected and susceptible persons. As such we again get (26), where R is the total number of persons in the system.

It may be noted that in both the examples, while $N(t)$ is essentially an integer-valued variable, we have treated it as a continuous variable. This can be regarded as an idealisation of the situation or as an approximation to reality.

2.3.3 Rate of Dissolution

Let $x(t)$ be the amount of undissolved solute in a solvent at time t and let c_0 be the maximum concentration or saturation concentration, i.e. the maximum amount of the solute that can be dissolved in a unit volume of the solvent. Let V be the volume of the solvent. It is found that the rate at which the solute is dissolved is proportional to the amount of undissolved solute and to the difference between the concentration of the solute at time t and the maximum possible concentration, so that we get

$$\frac{dx}{dt} = kx(t)\left(\frac{x(0) - x(t)}{V} - c_0\right) = \frac{kx(t)}{V}((x_0 - c_0 V) - x(t)) \quad (27)$$

2.3.4 Law of Mass Action: Chemical Reactions

Two chemical substances combine in the ratio $a : b$ to form a third substance Z. If $z(t)$ is the amount of the third substance at time t, then a proportion $az(t)/(a + b)$ of it consists of the first substance and a proportion $bz(t)/(a + b)$ of it consists of the second substance. The rate of formation of the third substance is proportional to the product of the amount of the two component substances which have not yet combined together. If A and B are the initial amounts of the two substances, then we get

$$\frac{dz}{dt} = k\left(A - \frac{az}{a + b}\right)\left(B - \frac{bz}{a + b}\right) \quad (28)$$

This is the non-linear differential equation for a second order reaction. Similarly for an nth order reaction, we get the non-linear equation

$$\frac{dz}{dt} = k(A_1 - a_1 z)(A_2 - a_2 z) \dots (A_n - a_n z), \quad (29)$$

where $a_1 + a_2 + \dots + a_n = 1$.

EXERCISE 2.3

1. If in (24), $a = 0.03134$, $b = (1.5887)(10)^{-10}$, $x(0) = 39 \times 10^6$, show that

$$x(t) = \frac{313,400,000}{1.5887 + 78,7703^{-0.03134t}}$$

This is Verhulst model for the population of USA when time zero corresponds to 1790. Estimate the population of USA in 1800, 1850, 1900 and 1950. Show that the point of inflexion should have occurred in about 1914. Find also the limiting population of USA on the basis of this model.

2. In (26) $k = 0.007$, $R = 1000$, $N(0) = 50$, find $N(10)$ and find when $N(t) = 500$.

3. Obtain the solution of (27) when $x_0 > c_0 V$ and $x_0 < c_0 V$ and interpret your results.

4. Obtain the solutions of (28) and (29).

5. Substances X and Y combine in the ratio 2 : 3 to form Z. When 45 grams of X and 60 grams of Y are mixed together, 50 gms of Z are formed in 5 minutes. How many grams of Z will be found in 210 minutes? How much time will it like to get 70 gms of Z?

6. Cigarette consumption in a country increased from 50 per capita in 1900 AD to 3900 per capita in 1960 AD. Assuming that the growth in consumption follows a logistic law with a limiting consumption of 4000 per capita, estimate the consumption per capita in 1950.

7. One possible weakness of the logistic model is that the average growth rate $1/x \, dx/dt$ is largest when x is small. Actually some species may become extinct if this population becomes very small. Suppose m is the minimum viable population for such a species, then show that

$$dx/dt = rx\left(1 - \frac{x}{k}\right)\left(1 - \frac{m}{x}\right)$$

has the desired property that x becomes extinct if $x_0 < m$. Also solve the differential equations in the two cases when $x_0 > m$ and $x_0 < m$.

8. Show that the logistic model can be written as

$$\frac{1}{N}\frac{dN}{dt} = r\left(\frac{K-N}{K}\right)$$

Deduce that K is the limiting size of the population and the average rate of growth is proportional to the fraction by which the population is unsaturated.

9. If $F(t)$ is the food consumed by population $N(t)$ and S is the food consumed by the population K, Smith replaced $(K-N)/N$ in Ex. 8 by $(S-F)/S$. He also argued that since a growing population consumes food faster than a saturated population, we should take $F(t) = c_1 N + c_2 \, dN/dt, c_1, c_2 > 0$. Use this assumption to modify the logistic model and solve the resulting differential equation.

9. A generalisation of the logistic model is

$$\frac{1}{N}\frac{dN}{dt} = \frac{r}{\alpha}\left(1 - \left(\frac{N}{K}\right)^\alpha\right), \quad \alpha > 0$$

Solve this differential equation. Show that the limiting population is still K and the point of inflexion occurs when the population in $K(\alpha+1)^{1/2\alpha}$. Show that this increases monotonically from $K/2$ to K as α increases from unity to ∞. What is the model if $\alpha \to 0$? What happens if $\alpha \to -1$?

10. A fish population which is growing according to logistic law is harvested at a constant rate H. Show that

$$\frac{dN}{dt} = rN\left(1 - \frac{N}{K}\right) - H$$

Show that if $D = kH/r - K^2/4 = a^2 > 0$, $N(t)$ approaches a constant limit as $t \to \pi/2 \, K/r^2$, but is discontinuous there and cannot predict beyond this value of t. If $D = 0$, show that the limiting population is $K/2$. If $D < 0$, show that the ultimate population size is $K/2(1 + \sqrt{1 - 4H/rK})$.

11. For each of the models discussed in this subsection, state explicitly the assumptions made. Try to extend the model when one or more of these assumptions are given up or modified. Obtain some critical results which may be different between the original and modified models and which may be capable of being tested through observations and experiments.

2.4 COMPARTMENT MODELS

In the last two sections, we got mathematical models in terms of ordinary differential equations of the first order, in all of which variables were separable. In the present section, we get models in terms of linear differential equations of first order.

We also use here the principle of continuity i.e. that the gain in amount of a substance in a medium in any time is equal to the excess of the amount that has entered the medium in the time over the amount that has left the medium in this time.

2.4.1 A Simple Compartment Model

Let a vessel contain a volume V of a solution with concentration $c(t)$ of a substance at time t (Figure 2.3) Let a solution with constant concentration C in an overhead tank enter the vessel at a constant rate R and after mixing thoroughly with the solution in the vessel, let the mixture with concentration $c(t)$ leave the vessel at the same rate R so that the volume of the solution in the vessel remains V.

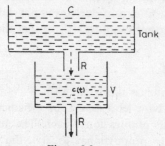

Figure 2.3

Using the principle of continuity, we get

$$V(c(t + \Delta t) - c(t)) = RC\Delta t - Rc(t)\Delta t + 0(\Delta t)$$

giving

$$V\frac{dc}{dt} + Rc = RC \tag{30}$$

Integrating

$$c(t) = c(0) \exp\left(-\frac{R}{V}t\right) + C\left(1 - \exp\left(-\frac{R}{V}t\right)\right) \tag{31}$$

As $t \to \infty$, $c(t) \to C$, so that ultimately the vessel has the same concentration as the overhead tank. Since

$$c(t) = C - (C - c_0) \exp\left(-\frac{R}{V}t\right), \tag{32}$$

if $C > c_0$, the concentration in the vessel increases to C; on the other hand if $C < c_0$, the concentration in the vessel decreases to C (Figure 2.4).

If the rate R' at which the solution leaves the vessel is less than R, the equations of continuity gives

$$\frac{d}{dt}[(V_0 + (R - R')t)c(t)]$$

$$= RC - R'(ct) \qquad (33)$$

where V is the initial volume of the solution in the vessel. This is also a linear differential equation of the first order.

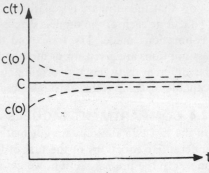

Figure 2.4

2.4.2 Diffusion of Glucose or a Medicine in the Blood Stream

Let the volume of blood in the human body be V and let the initial concentration of glucose in the blood stream be $c(0)$. Let glucose be introduced in the blood stream at a constant rate I. Glucose is also removed from the blood stream due to the physiological needs of the human body at a rate proportional to $c(t)$, so that the continuity principle gives

$$V\frac{dc}{dt} = I - kc \qquad (34)$$

which is similar to (30).

Now let a dose D of a medicine be given to a patient at regular intervals of duration T each. The medicine also disappears from the system at a rate proportional to $c(t)$, the concentration of the medicine in the blood stream, then the differential equation given by the continuity principle is

$$V\frac{dc}{dt} = -kc \qquad (35)$$

Integrating

$$c(t) = D \exp\left(-\frac{k}{V}t\right), \quad 0 \leqslant t < T \qquad (36)$$

At time T, the residue of the first dose is $D \exp\left(-\frac{k}{V}T\right)$ and now another dose D is given so that we get

$$c(t) = \left(D \exp\left(-\frac{k}{V}T\right) + D\right) \exp\left(-\frac{k}{V}(t - T)\right), \qquad (37)$$

$$= D \exp\left(-\frac{k}{V}t\right) + D \exp\left(-\frac{k}{V}(t - T)\right), \qquad (38)$$

$$T \leqslant t < 2T$$

The first term gives the residual of the first dose and the second term gives the residual of the second dose. Proceeding in the same way, we get after n doses have been given

$$c(t) = D \exp\left(-\frac{k}{V}t\right) + D \exp\left(-\frac{k}{V}(t - T)\right)$$

$$+ D \exp\left(-\frac{k}{V}(t - 2T)\right) + \ldots + D \exp\left(-\frac{k}{V}(t - \overline{n-1}T)\right) \tag{39}$$

$$= D \exp\left(-\frac{k}{V}t\right)\left(1 + \exp\left(\frac{k}{V}T\right) + \exp\left(\frac{2k}{V}T\right)\right.$$

$$\left. + \ldots + \exp\left((n-1)\frac{k}{V}T\right)\right)$$

$$= D \exp\left(-\frac{k}{V}t\right)\frac{\exp\left(n\frac{k}{V}T\right) - 1}{\exp\left(\frac{k}{V}T\right) - 1}, \quad (n-1)T \leqslant t < nT \tag{40}$$

$$c(nT - 0) = D \frac{1 - \exp\left(-\frac{k}{V}nT\right)}{\exp\left(\frac{kT}{V}\right) - 1} \tag{41}$$

$$c(nT + 0) = D \frac{\exp\left(\frac{kT}{V}\right) - \exp\left(-\frac{k}{V}nT\right)}{\exp\left(\frac{kT}{V}\right) - 1} \tag{42}$$

Thus the concentration never exceeds $D/\left(1 - \exp\left(-\frac{kT}{V}\right)\right)$. The graph of $c(t)$ is shown in Figure 2.5.

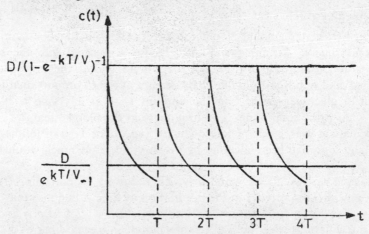

Figure 2.5

Thus in each interval, concentration decreases. In any interval, the concentration is maximum at the beginning of this interval and thus maximum concentration at the beginning of an interval goes on increasing as the number of intervals increases, but the maximum value is always below $D/(1 - e^{-kT/V})$. The minimum value in an interval occurs at the end of each interval. This also increases, but it lies below $D/(\exp(kT/V) - 1)$.

The concentration curve is piecewise continuous and has points of discontinuity at $T, 2T, 3T, \ldots$

By injecting glucose or penicillin in blood and fitting curve (36) to the data, we can estimate the value of k and V. In particular this gives a method for finding the volume of blood in the human body.

2.4.3 The Case of a Succession of Compartments

Let a solution with concentration $c(t)$ of a solute pass successively into n tanks in which the initial concentrations of the solution are $c_1(0), c_2(0), \ldots, c_n(0)$. The rates of inflow in each tank is the same as the rate of outflow from the tank. We have to find the concentrations $c_1(t), c_2(t) \ldots c_n(t)$ at time t. We get the equations

$$V\frac{dc_1}{dt} = Rc - Rc_1$$

$$V\frac{dc_2}{dt} = Rc_1 - Rc_2 \qquad\qquad (43)$$

$$\cdots\cdots\cdots\cdots$$

$$V\frac{dc_n}{dt} = Rc_{n-1} - Rc_n$$

By solving the first of these equations, we get $c_1(t)$. Substituting the value of $c_1(t)$ and proceeding in the same way, we can find $c_3(t), \ldots, c_n(t)$.

EXERCISE 2.4

1. Let $G(t)$ be the amount of glucose present in the blood-stream of a patient at time t. Assuming that the glucose is injected into the blood stream at a constant rate of C grames per minute, and at the same time is converted and removed from the blood stream at a rate proportional to the amount of glucose present, find the amount $G(t)$ at any time t. If $G(0) = G_0$, what is the equilibrium level of glucose in the blood stream?

2. A patient was given .5 micro-Curies (μc_i) of a type of iodine. Two hours later $.5\mu c_i$ had been taken up by his thryroid. How much would have been taken by the thyroid in two hours if he had been given $15\mu c_i$?

3. A gene has two alleles A and a which occur in proportions $p(t)$ and $q(t) = 1 - p(t)$ respectively in the population at time t. Suppose that allele A mutates to a at a constant rate μ. If $p(0) = q(0) = 1/2$, find $p(t)$ and $q(t)$. Write the equations when both alleles can mutate into each other at different rates.

4. A lake of constant volume V contains at time t an amount $Q(t)$ of pollutant evenly distributed throughout the lake. Suppose water containing concentration k of pollutant enters the lake at a rate r and water leaves the lake at the same rate. Suppose pollutants are also added to the lake at a constant rate P.

(a) If initial concentration of the pollutant in the lake is c_0, find $c(t)$.

(b) If there is no further addition of pollutant, in how many years, the pollutant concentration will be reduced to 10% of its present value?

(c) State explicitly the assumptions made in this model.

5. Suppose that a medicine disappears from the blood stream according to the law

$$\frac{dx}{dt} = -kx^2$$

and equal doses of this medicine are given at times $0, T, 2T, 3T, \ldots, nT, \ldots$. If x_n is the amount of the medicine in the blood stream immediately after the nth dose, show that the sequence $\{x_n\}$ is a monotonically increasing sequence. What is its limit as $n \to \infty$? Find the average amount of the medicine in the system in the time interval $(0, nT)$ and find the limit of this average amount as $n \to \infty$.

6. Repeat Exercise 5 for the law

$$\frac{dx}{dt} = -kx^m$$

7. Suppose in the model of subsection 2.4.2, we give a dose only when the concentration of the medicine in the blood stream falls to a prescribed level $D_0(< D)$. Find times $T_1, T_2, \ldots$ at which doses have to be given and discuss the behaviour of the sequence $\{T_n\}$ as n increases.

8. Compare the average concentration of a medicine in the system when (i) doses D are given at time intervals T and when (ii) doses $2D$ are given at time intervals $2T$.

2.5 MATHEMATICAL MODELLING IN DYNAMICS THROUGH ORDINARY DIFFERENTIAL EQUATIONS OF FIRST ORDER

Let a particle travel a distance x in time t in a straight line, then its velocity v is given by dx/dt and its acceleration is given by

$$dv/dt = (dv/dx)(dx/dt) = v\,dv/dx = d^2x/dt^2$$

2.5.1 Simple Harmonic Motion

Here a particle moves in a straight line in such a manner that its acceleration is always proportional to its distance from the origin and is always directed towards the origin, so that

$$v\frac{dv}{dx} = -\mu x \qquad (44)$$

Integrating

$$v^2 = \mu(a^2 - x^2), \qquad (45)$$

where the particle is initially at rest at $x = a$. Equation (44) gives

$$\frac{dx}{dt} = -\sqrt{\mu}\,\sqrt{a^2 - x^2} \qquad (46)$$

We take the negative sign since velocity increases as x decreases (Figure 2.6).

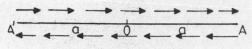

Figure 2.6

Integrating again and using the condition that at $t = 0$, $x = a$

$$x(t) = a \cos \sqrt{\mu}\, t \qquad (47)$$

so that

$$v(t) = -a\sqrt{\mu}\, \sin \sqrt{\mu}\, t, \qquad (48)$$

Thus in simple harmonic motion, both displacement and velocity are periodic functions with period $2\pi/\sqrt{\mu}$.

The particle starts from A with zero velocity and moves towards 0 with increasing velocity and reaches 0 at time $\pi/2\sqrt{\mu}$ with velocity $\sqrt{\mu}a$. It continue to move in the same direction, but now with decreasing velocity till it reaches $A'(0A' = a)$ where its velocity is again zero. It then begins moving towards 0 with increasing velocity and reaches 0 with velocity $\sqrt{\mu}a$ and again comes to rest at A after a total time period $2\pi/\sqrt{\mu}$. The periodic motion then repeats itself.

As one example of SHM, consider a particle of mass m attached to one end of a perfectly elastic string, the other end of which is attached to a fixed point 0 (Figure 2.7). The particle moves under gravity in vacuum.

Let l_0 be the natural length of the string and let a be its extension when the particle is in equilibrium so that by Hooke's law

$$mg = T_0 = \lambda \frac{a}{l_0} \qquad (49)$$

where λ is the coefficient of elasticity. Now let the string be further stretched a distance c and then the mass be left free. The equation of motion which states that mass $\times$ acceleration in any direction = force

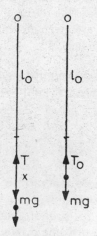

Figure 2.7

on the particle in that direction, gives

$$mv \frac{dv}{dx} = mg - T = mg - \lambda \frac{a + x}{l_0} = -\frac{\lambda x}{l_0} \qquad (50)$$

or

$$v \frac{dv}{dx} = \frac{\lambda}{m} \frac{x}{l_0} = -\frac{gx}{a}, \qquad (51)$$

which gives a simple harmonic motion with time period $2\pi \sqrt{\frac{a}{g}}$.

2.5.2 Motion Under Gravity in a Resisting Medium

A particle falls under gravity in a medium in which the resistance is proportional to the velocity. The equation of motion is

$$m \frac{dv}{di} = mg - mkv$$

or

$$\frac{dv}{V - v} = k \, dt; \quad V = \frac{g}{k} \qquad (52)$$

Integrating

$$V - v = Ve^{-kt} \qquad (53)$$

If the particle starts from rest with zero velocity. Equation (50) gives

$$v = V(1 - e^{-kt}), \qquad (54)$$

so that the velocity goes on increasing and approaches the limiting velocity g/k as $t \to \infty$. Replacing v by dx/dt, we get

$$\frac{dx}{dt} = V(1 - e^{-kt}) \qquad (55)$$

Integrating and using $x = 0$ when $t = 0$, we get

$$x = Vt + \frac{Ve^{-kt}}{k} - \frac{V}{k}. \qquad (56)$$

2.5.3 Motion of a Rocket

As a first idealisation, we neglect both gravity and air resistance. A rocket moves forward because of the large supersonic velocity with which gases produced by the burning of the fuel inside the rocket come out of the converging-diverging nozzle of the rocket (Figure 2.8).

Let $m(t)$ be the mass of the rocket at time t and let it move forward with velocity $v(t)$ so that the momentum at time t is $m(t)v(t)$.

In the interval of time $(t, t + \Delta t)$, the mass of the rocket becomes

$$m(t + \Delta t) = m(t) + \frac{dm}{dt} \Delta t + 0(\Delta t)$$

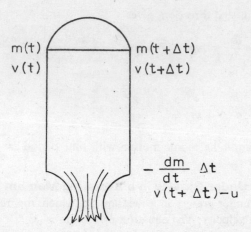

Figure 2.8

Since the rocket is losing mass, dm/dt is negative and the mass of gases $-dm/dt\ \Delta t$ moves with velocity u relative to the rocket, i.e. with a velocity $v(t + \Delta t) - u$ relative to the earth so that the total momentum of the rocket and the gases at time $t + \Delta t$ is

$$m(t + \Delta t)v(t + \Delta t) - \frac{dm}{dt}\Delta t(v(t + \Delta t) - u) \qquad (57)$$

Since we are neglecting air resistance and gravity, there is no external force on the rocket and as such the momentum is conserved, giving the equation

$$m(t)v(t) = \left(m(t) + \frac{dm}{dt}\Delta t\right)\left(v(t) + \frac{dv}{dt}\Delta t\right)$$

$$- \frac{dm}{dt}\Delta t(v - u) + 0(\Delta t)^2 \qquad (58)$$

Dividing by Δt and proceeding to the limit as $\Delta t \to 0$, we get

$$m(t)\frac{dv}{dt} = -u\frac{dm}{dt} \qquad (59)$$

or

$$\frac{dm}{m} = -\frac{1}{u}dv \qquad (60)$$

or

$$\ln\frac{m(t)}{m(0)} = -\frac{v(t)}{u} \qquad (61)$$

assuming that the rocket starts with zero velocity.

As the fuel burns, the mass of the rocket decreases. Initially the mass of the rocket $= m_P + m_F + m_S$ when m_P is the mass of the pay-load, m_F is the mass of the fuel and m_S is the mass of the structure. When the fuel is completely burnt out, m_F becomes zero and if v_B is the velocity of the rocket at this stage, when the fuel is all burnt, then (60) gives

$$v_B = u \ln \frac{m_P + m_F + m_S}{m_P + m_S} = u \ln \left(1 + \frac{m_F}{m_P + m_S}\right) \tag{62}$$

This is the maximum velocity that the rocket can attain and it depends on the velocity u of efflux of gases and the ratio $m_F/(m_P + m_S)$. The larger the values of u and $m_F/(m_P + m_S)$, the larger will be the maximum velocity attained.

For the best modern fuels and structural materials, the maximum velocity this gives is about 7 km/sec. In practice it would be much less since we have neglected air resistance and gravity, both of which tend to reduce the velocity. However if a rocket is to place a satellite in orbit, we require a velocity of more than 7 km/sec.

The problem can be overcome by using the concept of multi-stage rockets.

The fuel may be carried in a number of containers and when the fuel of a container is burnt up, the container is thrown away, so that the rocket has not to carry any dead weight.

Thus in a three-stage rocket, let m_{F_1}, m_{F_2}, m_{F_3} be the masses of the fuels and m_{S_1}, m_{S_2}, m_{S_3} be the three corresponding masses of containers, then velocity at the end of the first stage is

$$v_1 = u \ln \frac{m_P + m_{F_1} + m_{S_1} + m_{F_2} + m_{S_2} + m_{F_3} + m_{S_3}}{m_P + m_{F_2} + m_{S_2} + m_{F_3} + m_{S_3}} \tag{63}$$

At the end the second stage, the velocity is

$$v_2 = v_1 + u \ln \frac{m_P + m_{F_2} + m_{F_3} + m_{S_3}}{m_P + m_{F_3} + m_{S_3}} \tag{64}$$

and at the end of the third stage, the velocity

$$v_3 = v_2 + u \ln \frac{m_P + m_{F_3}}{m_P} \tag{65}$$

In this way, a much larger velocity is obtained than can be obtained by a single-stage rocket.

EXERCISE 2.5

1. Discuss the problem of Section 2.5.1 when the particle start from A with velocity v_0 away from the origin.

2. Draw the graphs of $v(t)$ and $x(t)$ against t for two complete oscillations.

3. Discuss the motion of the particle in Section 2.5.2 when $c > a$.

4. Show that for the same pay-load, same total fuel mass and some total structure mass, the final velocity of a multistage rocket is more than that of a single-stage rocket.

5. Discuss the motion of a rocket when gravity is taken into account.

6. If the particle attached to the elastic string in Figure 2.7 moves in a resisting medium, discuss its motion when the resistance is proportional to (i) velocity (ii) square of the velocity.

7. Discuss the motion of a particle projected vertically upwards under gravity with initial velocity U when the air resistance is proportional to the square of the velocity. With what velocity will the particle return to the Earth?

8. Assuming that a particle projected vertically upwards from the surface of the earth moves in vacuum under a force ga^2/x^2 directed toward the centre of earth, where x is the distance of the particle from the centre of the earth, find the initial velocity of projection so that the particle never return to earth.

2.6 MATHEMATICAL MODELLING OF GEOMETRICAL PROBLEMS THROUGH ORDINARY DIFFERENTIAL EQUATIONS OF FIRST ORDER

2.6.1 Simple Geometrical Problems

Many geometrical entities can be expressed in terms of derivatives and as such relations between these entities can give rise to differential equations whose solution will give us a family of curves for which the given relation between geometrical entities is satisfied.

(i) Find curves for which tangent at a point is always perpendicular to the line joining the point to the origin.

The slope of the tangent is dy/dx and the slope of line joining the point (x, y) to the origin is y/x and since these lines are given to be orthogonal

$$\frac{dy}{dx} = -\frac{x}{y} \tag{66}$$

Integrating

$$x^2 + y^2 = a^2 \tag{67}$$

which represents a family of concentric circle.

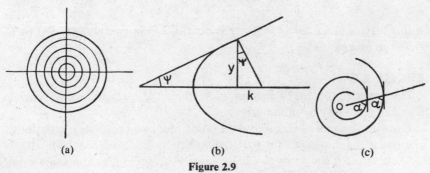

(a)   (b)   (c)

Figure 2.9

(ii) Find curves for which the projection of the normal on the x-axis is of constant length.

This condition gives

$$y \frac{dy}{dx} = k \tag{68}$$

Integrating

$$y^2 = 2kx + A, \tag{69}$$

which represents a family of parabolas, all with the same axis and same length of latus rectum.

(iii) Find curves for which tangent makes a constant angle with the radius vector.

Here it is convenient to use polar coordinates and the conditions of the problem gives

$$r \frac{d\theta}{dr} = \tan \alpha \tag{70}$$

Integrating

$$r = Ae^{\theta \cot \alpha}, \tag{71}$$

which represents a family of equiangular spirals.

2.6.2 Orthogonal Trajectories

Let

$$f(x, y, a) = 0 \tag{72}$$

represent a family of curves, one curve for each value of the parameter a. Differentiating (72), we get

$$\frac{\partial f}{\partial x} + \frac{\partial f}{\partial y} \frac{dy}{dx} = 0 \tag{73}$$

Eliminating a between (72) and (73), we get a differential equation of the first order

$$\varphi \left(x, y, \frac{dy}{dx} \right) = 0, \tag{74}$$

of which (72) is the general solution. Now we want a family of curves cutting every member of (72) at right angle at all points of intersection.

At a point of intersection of the two curves, x, y are the same but the slope of the second curve is negative reciprocal of the slope of the first curve. As such differential equation of the family of orthogonal trajectories is

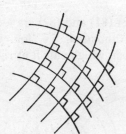

Figure 2.10

$$\varphi \left(x, y, - \frac{1}{dy/dx} \right) = 0 \tag{75}$$

Integrating (75), we get

$$g(x, y, b) = 0, \tag{76}$$

which give the orthogonal trajectories of the family (72).

(i) Let the original family be $y = mx$, when m is a parameter then

$$dy/dx = m$$

and eliminating m, we get the differential equation of this concurrent family of straight lines as

$$\frac{y}{x} = \frac{dy}{dx} \tag{77}$$

To get the orthogonal trajectories, we replace dy/dx by $-1/(dy/dx)$ to get

$$\frac{y}{x} = -\frac{1}{dy/dx}$$

Integrating

$$x^2 + y^2 = a^2 \tag{78}$$

which gives the orthogonal trajectories as concentric circles (Figure 2.9a).

(i) Find the orthogonal trajectories of the family of confocal conics

$$\frac{x^2}{a^2 + \lambda} + \frac{y^2}{b^2 + \lambda} = 1, \tag{79}$$

where λ is a parameter. Differentiating, we get

$$\frac{x}{a^2 + \lambda} + \frac{y}{b^2 + \lambda}\frac{dy}{dx} = 0 \tag{80}$$

Eliminating λ between (79) and (80), we get

$$(xp - y)(x + py) = p(a^2 - b^2); \quad p = \frac{dy}{dx} \tag{81}$$

To get the orthogonal trajectories, we replace p by $-\dfrac{1}{p}$ to get

$$\left(-\frac{x}{p} - y\right)\left(x - \frac{y}{p}\right) = -\frac{1}{p}(a^2 - b^2)$$

or
$$(xp - y)(x + py) = p(a^2 - b^2) \tag{82}$$

However (81) and (82) are identical. As such the family of confocal conics is self-orthogonal, i.e. for every conic of the family, there is another with same focii which cuts it at right angles.

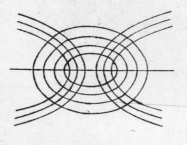

One family consists of confocal ellipses and the other consists of confocal hyperbolas with the same focii (Figure 2.11).

(iii) In polar coordinates after getting the differential equation of the family of curves, we have to replace $r\dfrac{d\theta}{dr}$ by $-1\Big/\Big(r\dfrac{d\theta}{dr}\Big)$ and then integrate the resulting differential equation.

Figure 2.11

Then if the original family is

$$r = 2a \cos \theta, \tag{83}$$

with $a > 0$ as a parameter, its differential equation is obtained by eliminating a between (83) and

$$\frac{dr}{d\theta} = -2a \sin \theta \tag{84}$$

to get

$$r\frac{d\theta}{dr} = -\cot \theta \tag{85}$$

Replacing $r\frac{d\theta}{dr}$ by $-\left(r\frac{d\theta}{dr}\right)^{-1}$, we get

$$r\frac{d\theta}{dr} = \tan \theta \tag{86}$$

Integrating we get

$$r = 2b \sin \theta \tag{87}$$

The orthogonal trajectories are shown in Figure 2.12.

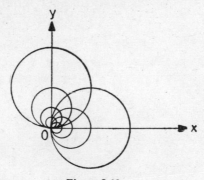

Figure 2.12

The circles of both families pass through the origin, but while the centre of one family lie on x-axis, the centres of the orthogonal family lie on y-axis.

EXERCISE 2.6

1. Find a family of curves such that for each curve, the length of the tangent intercepted between the axes is of constant length. Draw the curves.

2. Find a family of curves such that for each curve, the length of tangent intercepted between the point (x, y) and the axis of y is of constant length.

3. Find a curve such that all rays of light starting from the origin are reflected from points of the curve in the direction of the y-axis.

4. Find a curve such that all rays emanating from a given point $(-a, 0)$ after being reflected from points on the curve pass through the point $(a, 0)$.

5. Find the orthogonal trajectories of the families of curves

(i) $y^2 = 4cx$ (ii) $x^2 + y^2 - 2ax = 0$

(iii) $r = ae^{\theta \cot \alpha}$ (iv) $y^2 = 4cx + 4c^2$

(v) $r = a(1 + \cos \theta)$

5. In electrostatics, lines of force always cut equipotential curves (surfaces) at right angles. Find lines of force and equipotential surfaces for (i) one charge (ii) for two charges, and verify the result stated.

Mathematical Modelling Through Systems of Ordinary Differential Equations of the First Order

3.1 MATHEMATICAL MODELLING IN POPULATION DYNAMICS

3.1.1 Prey-Predator Models

Let $x(t)$, $y(t)$ be the populations of the prey and predator species at time t. We assume that

(i) if there are no predators, the prey species will grow at a rate proportional to the population of the prey species,

(ii) if there are no prey, the predator species will decline at a rate proportional to the population of the predator species,

(iii) the presence of both predators and preys is beneficial to growth of predator species and is harmful to growth of prey species. More specifically the predator species increases and the prey species decreases at rates proportional to the product of the two populations.

These assumptions give the systems of non-linear first order ordinary differential equations

$$\frac{dx}{dt} = ax - bxy = x(a - by), \qquad a, b > 0 \tag{1}$$

$$\frac{dy}{dt} = -py + qxy = -y(p - qx), \quad p, q > 0 \tag{2}$$

Now dx/dt, dy/dt both vanish if

$$x = x_e = \frac{p}{q}, \qquad y = y_e = \frac{a}{b}. \tag{3}$$

If the initial populations of prey and predator species are p/q and a/b respectively, the populations will not change with time. These are the equilibrium sizes of the populations of the two species. Of course $x = 0$, $y = 0$ also gives another equilibrium position.

From (1) and (2)

$$\frac{dy}{dx} = -\frac{y(p - qx)}{x(a - by)} \tag{4}$$

or

$$\frac{a - by}{y} dy = -\frac{p - qx}{x} dx; \quad x_0 = x(0), \quad y_0 = y(0) \tag{5}$$

Integrating

$$a \ln \frac{y}{y_0} + p \ln \frac{x}{x_0} = b(y - y_0) + q(x - x_0) \tag{6}$$

Thus through every point of the first quadrant of the x–y plane, there is a unique trajectory. No two trajectories can intersect, since intersection will imply two different slopes at the same point.

If we start with $(0, 0)$ or $(p/q, a/b)$, we get point trajectories. If we start with $x = x_0, y = 0$, from (1) and (2), we find that x increases while y remains zero. Similarly if we start with $x = 0$, $y = y_0$, we find that x remains zero while y decreases. Thus positive axes of x and y give two line trajectories (Figure 3.1).

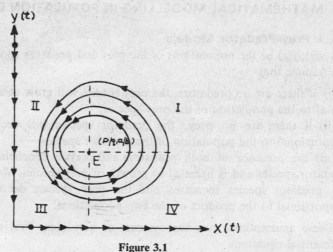

Figure 3.1

Since no two trajectories intersect, no trajectory starting from a point situated within the first quadrant will intersect the x-axis and y-axis trajectories. Thus all trajectories corresponding to positive initial populations will lie strictly within the first quadrant. Thus if the initial populations are positive, the populations will be always positive. If the population of one (or both) species is initially zero, it will always remain zero.

The lines through $(p/q, a/b)$ parallel to the axes of coordinates divide the first quadrant into four parts I, II, III and IV. Using (1), (2), we find that

$$\text{in I,} \quad dx/dt < 0, \quad dy/dt > 0, \quad dy/dx < 0$$
$$\text{in II,} \quad dx/dt < 0, \quad dy/dt < 0, \quad dy/dx > 0$$
$$\text{in III,} \quad dx/dt > 0, \quad dy/dt < 0, \quad dy/dx < 0$$
$$\text{in IV,} \quad dx/dt > 0, \quad dy/dt > 0, \quad dy/dx > 0$$

This give the direction field at all points as shown in Figure 3.1. Each trajectory is a closed convex curve. These trajectories appear relatively cramped near the axes.

In I and II, prey species decreases and in III and IV, it increases. Similarly in IV and I, predator species increases and in II and III, it decreases. After a certain period, both species return to their original sizes and thus both species sizes vary periodically with time.

3.1.2 Competition Models

Let $x(t)$ and $y(t)$ be the populations of two species competing for the same resources, then each species grows in the absence of the other species, and the rate of growth of each species decreases due to the presence of the other species. This gives the system of differential equations

$$\frac{dx}{dt} = ax - bxy = bx\left(\frac{a}{b} - y\right); \quad a > 0, \quad b > 0 \tag{7}$$

$$\frac{dy}{dt} = py - qxy = y(p - qx) = qy\left(\frac{p}{q} - x\right); \quad p > 0, \quad q > 0 \tag{8}$$

There are two equilibrium positions viz. $(0, 0)$ and $(p/q, a/b)$. There are two point trajectories viz. $(0, 0)$ and $(p/q, a/b)$ and there are two line trajectories viz. $x = 0$ and $y = 0$.

$$
\begin{array}{llll}
\text{In I} & dx/dt < 0, & dy/dt < 0, & dy/dx > 0 \\
\text{In II} & dx/dt < 0, & dy/dt > 0, & dy/dx < 0 \\
\text{In III} & dx/dt > 0, & dy/dt > 0, & dy/dx > 0 \\
\text{In IV} & dx/dt > 0, & dy/dt < 0, & dy/dx < 0
\end{array}
$$
(9)

(10)

This gives the direction field as shown in Figure 3.2. From (7) and (8)

$$\frac{dy}{dx} = \frac{y(p - qx)}{x(a - by)} \quad \text{or} \quad \frac{a - by}{y}\, dy = \frac{p - qx}{x}\, dx \tag{11}$$

Integrating

$$a \ln \frac{y}{y_0} - b(y - y_0) = p \ln \frac{x}{x_0} - q(x - x_0) \tag{12}$$

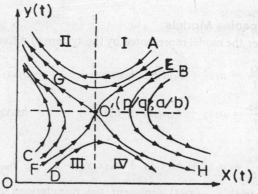

Figure 3.2

The trajectory which passes through $(p/q, a/b)$ is

$$a \ln \frac{by}{a} - by + a = p \ln \frac{qx}{p} - qx + p \tag{13}$$

If the initial populations correspond to the point A, ultimately the first species dies but and the second species increases in size to infinity. If the initial populations correspond to the point B, then ultimately the second species dies out and the first species tends to infinity. Similarly if the initial populations correspond to point C, the first species dies out and the second species goes to infinity and if the initial populations correspond to point D, the second species dies out and the first species goes to infinity.

If the initial populations correspond to point E or F, the species populations converge to equilibrium populations $p/q, a/b$ and if the initial population correspond to point G, H, the first and second species die out respectively.

Thus except when the initial populations correspond to points on curves $O'E$ and $O'F$, only one species will survive in the competition process and the species can coexist only when the initial population sizes correspond to points on the curve EF.

It is also interesting to note that while the initial populations corresponding to A, E, B are quite close to one another, the ultimate behaviour of these populations are drastically different. For populations starting at A, the second species alone survives, for populations starting at B, the first species alone survives, while for population starting at E, both species can coexist. Thus a slight change in the initial population sizes can have a catastrophic effect on the ultimate behaviour.

It may also be noted that for both prey-predator and competition models, we have obtained a great deal of insight into the models without using the solution of these equations (1), (2) or (7), (8). By using numerical methods of integration with the help of computers, we can draw some typical trajectories in both cases and can get additional insight into the behaviour of these models.

3.1.3 Multi-species Models

We can consider the model represented by the system of differential equations

$$\frac{dx_1}{dt} = a_1 x_1 + b_{11} x_1^2 + b_{12} x_1 x_2 + \ldots + b_{1n} x_1 x_n$$

$$\frac{dx_2}{dt} = a_2 x_2 + b_{21} x_2 x_1 + b_{22} x_2^2 + \ldots + b_{2n} x_2 x_n \tag{14}$$

$$\ldots \ldots \ldots \ldots \ldots \ldots \ldots \ldots \ldots \ldots$$

$$\frac{dx_n}{dt} = a_n x_n + b_{n1} x_n x_1 + b_{n2} x_n x_2 + \ldots + b_{nn} x_n^2$$

Here $x_1(t)$, $x_2(t)$, $\ldots$, $x_n(t)$ represent the populations of the n species. Also a_i is positive or negative according as the ith species grows or decays in the absence of other species and b_{ij} is positive or negative according as the ith species benefits or is harmed by the presence of the jth species. In general b_{ii} is negative since members of the ith species also compete among themselves for limited resources.

We can find the positions of equilibrium by putting

$$dx_i/dt = 0 \qquad \text{for} \qquad i = 1, 2, \ldots, n$$

and solving the n algebraic equations for $x_1, x_2, \ldots, x_h$. We can also obtain all degenerate solutions in which one or more x_i's are zero, i.e. in which one or more species have disappeared and finally we have the equilibrium position in which all species can disappear.

If $x_{10}, x_{20}, \ldots, x_{n0}$ is an equilibrium position, we can discuss its local stability by substituting

$$x_1 = x_{10} + u_1, \quad x_2 = x_{20} + u_2, \ldots, \qquad x_n = x_{n0} + u_n \tag{15}$$

14) and getting a system of linear differential equations

$$\frac{du_1}{dt} = c_{11}u_1 + c_{12}u_2 + \ldots + c_{1n}u_n$$

$$\frac{du_2}{dt} = c_{21}u_1 + c_{22}u_2 + \ldots + c_{2n}u_n \tag{16}$$

$$\cdots\cdots\cdots\cdots\cdots\cdots\cdots\cdots\cdots\cdots$$

$$\frac{du_n}{dt} = c_{n1}u_1 + c_{n2}u_2 + \ldots + c_{nn}u_n,$$

by neglecting squares, products and higher powers of u_i's. We can try the solutions $u_1 = A_1 e^{\lambda t}, u_2 = A_2 e^{\lambda t}, \ldots, u_n = A_n e^{\lambda t}$ to get

$$\begin{vmatrix} c_{11} - \lambda & c_{12} & c_{13} & \ldots & c_{1n} \\ c_{21} & c_{22} - \lambda & c_{23} & \ldots & c_{2n} \\ \cdots & \cdots & \cdots & \cdots & \cdots \\ c_{n1} & c_{n2} & c_{n3} & \ldots & c_{nn} - \lambda \end{vmatrix} = 0 \tag{17}$$

Thus the equilibrium position would be stable if the real parts of all the eigenvalues of the matrix $[c_{ij}]$ are negative. The conditions for this are given by Routh-Hurwitz criterion which states that all the roots of

$$a_0 x^n + a_1 x^{n-1} + \ldots + a_n = 0, \qquad a_0 > 0 \tag{18}$$

will have negative real parts if and only if $T_0, T_1, T_2, \ldots$ are positive where

$$T_0 = a_0, \quad T_1 = a_1, \quad T_2 = \begin{vmatrix} a_1 & a_0 \\ a_3 & a_2 \end{vmatrix}, \quad T_3 = \begin{vmatrix} a_1 & a_0 & 0 \\ a_3 & a_2 & a_1 \\ a_5 & a_4 & a_3 \end{vmatrix}$$

$$T_4 = \begin{vmatrix} a_1 & a_0 & 0 & 0 \\ a_3 & a_2 & a_1 & 0 \\ a_5 & a_4 & a_3 & a_2 \\ a_7 & a_6 & a_5 & a_4 \end{vmatrix} \tag{19}$$

This is true if and only if $a_i > 0$ and either all even-numbered T_k or all odd-numbered T_k are positive. Alternatively (18) will have all roots with negative real parts iff this is true for the $(n - 1)$th degree equation

$$a_1 x^{n-1} + a_2 x^{n-2} + a_3 x^{n-3} + \ldots - \frac{a_0}{a_1} a_3 x^{n-2} - \frac{a_0}{a_1} a_5 x^{n-4} - \ldots = 0 \quad (20)$$

The above method will enable us to discuss only local stability of a position of equilibrium, i.e. this will decide that if the populations of different species are changed slightly from these equilibrium values, whether the population sizes will return to their original equilibrium values or not. The problem of discussing the global stability i.e. of discussing whether the populations will return to these equilibrium values, whatever be the magnitudes of the disturbances, is a more difficult problem and it is possible to solve this problem in special cases only.

3.1.4 Age-Structured Population Models

Let $x_1(t)$, $x_2(t)$, . . . , $x_p(t)$ be the populations of the p pre-reproductive age-groups; let $x_{p+1}(t)$, . . . , $x_{p+q}(t)$ be the populations of q reproductive age-groups and let $x_{p+q+1}(t)$, . . . , $x_{p+q+r}(t)$ be the populations of the r post-reproductive age-groups. Let b_{p+1}, b_{p+2}, . . . , b_{p+q} be the birth rates in the q reproductive age-groups, let d_i be the death rates in the ith age-group $(i = 1, 2, \ldots, p + q + r)$ and let m_j be the rate of migration from the jth age-group to the $(j + 1)$th age-group $(j = 1, 2, \ldots, p + q + r - 1)$, then we get the system of differential equations

$$\frac{dx_1}{dt} = b_{p+1} x_{p+1} + \ldots + b_{p+q} x_{p+q} - (d_1 + m_1) x_1$$

$$(21)$$

$$\frac{dx_2}{dt} = m_1 x_1 - (d_2 + m_2) x_2$$

. . . .

$$\frac{dx_n}{dt} = m_{n-1} x_{n-1} - d_n x_n; \quad n = p + q + r$$

or

$$\frac{d}{dt}
\begin{bmatrix}
x_1(t) \\
x_2(t) \\
\cdot \\
\cdot \\
\cdot \\
x_n(t)
\end{bmatrix}$$

$$=
\begin{bmatrix}
-(d_1 + m_1) & 0 & .. & b_{p+1} & \cdots & b_{p+q} & \cdots & 0 & 0 \\
m_1 & -(d_2 + m_2) & .. & 0 & \cdots & 0 & \cdots & 0 & 0 \\
0 & m_2 & .. & 0 & \cdots & 0 & \cdots & 0 & 0 \\
.. & & .. & .. & & .. & \cdots & .. & .. \\
.. & & .. & .. & .. & .. & \cdots & 0 & \cdots & m_{n-1} & -d_n
\end{bmatrix}$$

$$\times \begin{bmatrix} x_1(t) \\ x_2(t) \\ \cdot \\ \cdot \\ \cdot \\ x_n(t) \end{bmatrix} \qquad (22)$$

or

$$\frac{dX}{dt} = AX(t), \qquad (23)$$

where A is a matrix, all of whose diagonal elements are negative, all of whose main subdiagonal elements are positive, q other elements of the first row are positive and all other elements are zero. Equation (22) has the solution

$$X(t) = \exp{(At)}X(0) \qquad (24)$$

EXERCISES 3.1

1. Draw some trajectories for the model

$$\frac{dx}{dt} = x(1 - 0.1y), \qquad \frac{dy}{dt} = -y(1 - 0.1x)$$

2 Discuss the stability of the equilibrium positions $(0, 0)$ and $(p/q, a/b)$ for the prey-predator model represented by equations (1) and (2) and the competition model represented by equations (7) and (8).

3. Draw some trajectories for the competition model

$$\frac{dx}{dt} = x(1 - 0.1y), \qquad \frac{dy}{dt} = y(1 - 0.1x).$$

4. By integrating (1), (2) round a closed trajectory, show that

$$0 = a\bar{x} - b\overline{xy}, \qquad 0 = -p\bar{y} + q\overline{xy}$$
$$0 = a - b\bar{y}, \qquad 0 = -p + q\bar{x},$$

where $\quad \bar{x} = \frac{1}{T}\int_0^T x(t)\,dt, \quad \bar{y} = \frac{1}{T}\int_0^T y(t)\,dt, \quad \overline{xy} = \frac{1}{T}\int_0^T x(t)y(t)\,dt,$

and T is the time for the populations to return to original values.

5. Write the basic equations for the wolf-goat-cabbage model in which wolves eat goats, goats eat cabbages, but wolves do not eat cabbages.

6. Show that the model represented by

$$\frac{dx}{dt} = x(4 - x - y), \qquad \frac{dy}{dt} = y(15 - 5x - 3y), \ x \geqslant 0, \quad y \geqslant 0$$

has a position of equilibrium, this position is stable and two species can coexist.

7. Show that the model represented by

$$\frac{dx}{dt} = x(15 - 5x - 3y), \qquad \frac{dy}{dt} = y(4 - x - y), \ x \geqslant 0, \qquad y \geqslant 0$$

has a position of equilibrium, this position is unstable, only one species will survive and which species survives depends on initial conditions:

8. Show that the model represented by

$$\frac{dx}{dt} = x(30 - 60 - 5y), \frac{dy}{dt} = y(12 - 40 - 3y), \quad x \geqslant 0, \quad y \geqslant 0$$

has no position of equilibrium and that only the first species will survive.

9. Show that the model represented by

$$\frac{dx}{dt} = x(12 - 4x - 3y), \quad \frac{dy}{dt} = y(30 - 6x - 5y), \quad x \geqslant 0, y \geqslant 0$$

has no position of equilibrium and that only the second species will survive.

10. For the model representing competition between two species, each of which can exist and grow without the other and contact between which inhibits the growth of both, the differential equations are given by

$$\frac{dx}{dt} = x(A_1 - B_1 x - C_1 y), \quad \frac{dy}{dt} = y(A_2 - B_2 y - C_2 x),$$

where A_1, B_1, C_1, A_2, B_2, C_2 are all positive.

Show that

(i) the equilibrium will be biologically meaningful, i.e. the equilibrium position will be in the first quadrant if

$$B_2/C_2 > A_2/A_1 > C_2/B_1 \text{ or } C_2/B_1 > A_2/A_1 > B_2/C_1.$$

(ii) if a biologically meaningful equilibrium exists, it will be stable iff $B_1 B_2 > C_1 C_2$, i.e. if the product of self-restraint coefficients is greater than the product of the other restraint coefficients.

(iii) if the equilibrium does not exist, the first species will survive if

$$A_1/C_2 > A_2/B_2 \quad \text{and} \quad A_1/B_1 > A_2/C_2.$$

11. Discuss the modification of the prey-predator model when

(i) the predator population is harvested at a constant rate h_1 or
(ii) the prey population is harvested at a constant rate h_2 or
(iii) both species are harvested at constant rates.

12. Discuss the possibility of the existence of a stable age-structure i.e. age-structure which does not change with time in the model of Section 3.1.4.

3.2 MATHEMATICAL MODELLING OF EPIDEMICS THROUGH SYSTEMS OF ORDINARY DIFFERENTIAL EQUATIONS OF FIRST ORDER

3.2.1 A Simple Epidemic Model

Let $S(t)$ and $I(t)$ be the number of susceptibles (i.e. those who can get a disease) and infected persons (i.e. those who have already got the disease).

Initially let there be n susceptible and one infected person in the system so that

$$S(t) + I(t) = n + 1, \qquad S(0) = n, \qquad I(0) = 1 \tag{25}$$

The number of infected persons grows at a rate proportional to the product of susceptible and infected persons and the number of susceptible persons decreases at the same rate so that we get the system of differential equations

$$\frac{dS}{dt} = -\beta SI, \quad \frac{dI}{dt} = \beta SI, \tag{26}$$

so that

$$\frac{dS}{dt} + \frac{dI}{dt} = 0, \quad S(t) + I(t) = \text{constant} = n + 1 \tag{27}$$

and

$$\frac{dS}{dt} = -\beta S(n + 1 - S),$$

$$\frac{dI}{dt} = \beta I(n + 1 - I) \tag{28}$$

Integrating

$$S(t) = \frac{n(n + 1)}{n + e^{(n+1)\beta t}}, \quad I(t) = \frac{(n + 1)e^{(n+1)\beta t}}{n + e^{(n+1)\beta t}}, \tag{29}$$

so that

$$\underset{t \to \infty}{\text{Lt}} \ S(t) = 0, \qquad \underset{t \to \infty}{\text{Lt}} \ I(t) = n + 1 \tag{30}$$

3.2.2 A Susceptible-Infected-Susceptible (SIS) Model

Here, a susceptible person can become infected at a rate proportional to SI and an infected person can recover and become susceptible again at a rate γI, so that

$$\frac{dS}{dt} = -\beta SI + \gamma I, \quad \frac{dI}{dt} = \beta SI - \gamma I, \tag{31}$$

which gives

$$\frac{dI}{dt} = (\beta(n + 1) - \gamma)I - \beta I^2 \tag{32}$$

3.2.3 SIS Model with Constant Number of Carriers

Here infection is spread both by infectives and a constant number C of carriers, so that (30) becomes

$$\frac{dI}{dt} = \beta(I + C)S - \gamma I$$

$$= \beta C(n + 1) + \beta(n + 1 - C - \gamma/\beta)I - \beta I^2. \tag{33}$$

3.2.4 Simple Epidence Model with Carriers

In this model, only carriers spread the disease and their number decreases exponentially with time as these are identified and eliminated, so that we get

$$\frac{dS}{dt} = -\beta S(t)C(t) + \gamma I(t), \qquad \frac{dI}{dt} = \beta C(t)S(t) - \gamma I(t),$$

$$\frac{dC}{dt} = -\alpha C \tag{34}$$

so that

$$S(t) + I(t) = S_0 + I_0 = N \text{ (say)}, \quad C(t) = C_0 \exp(-\alpha t) \tag{35}$$

and

$$\frac{dI}{dt} = \beta C_0 N \exp(-\alpha t) - [\beta C_0 \exp(-\alpha t) + \gamma]I \tag{36}$$

3.2.3 Model with Removal

Here infected persons are removed by death or hospitalisation at a rate proportional to the number of infectives, so that the model is

$$\frac{dS}{dt} = -\beta SI, \qquad \frac{dI}{dt} = \beta SI - \gamma I = \beta I\left(S - \frac{\gamma}{\beta}\right)$$

$$= \beta I(S - \rho); \quad \rho = \frac{\gamma}{\beta} \tag{37}$$

with initial conditions

$$S(0) = S_0 > 0, \quad I(0) = I_0 > 0, \quad R(0) = R_0 = 0,$$

$$S_0 + I_0 = N. \tag{38}$$

3.2.6 Model with Removal and Immigration

We modify the above model to allow for the increase of susceptibles at a constant rate μ so that the model is

$$\frac{dS}{dt} = -\beta SI + \mu, \frac{dI}{dt} = \beta SI - \gamma I, \frac{dR}{dt} = \gamma I. \tag{39}$$

EXERCISE 3.2

1. Verify (29) and (30).
2. Integrate (32) and show that

$$\underset{t \to \infty}{\text{Lt}} I(t) = n + 1 - \rho \quad \text{if} \quad n + 1 > \rho = \gamma/\beta$$

$$= 0 \quad \text{if} \quad n + 1 \leqslant \rho = \gamma/\beta$$

3. Solve SIS model when β is a known function of t.
4. Integrate (36) and find limit of $I(t)$ as $t \to \infty$.
5. Discuss integration of models given by (37) and (39) and interpret your results.

3.3 COMPARTMENT MODELS THROUGH SYSTEMS OF ORDINARY DIFFERENTIAL EQUATIONS

Pharmokinetics (also called drug kinetics or tracer kinetics or multi-compartment analysis) deals with the distribution of drugs, chemicals, tracers or radio-active substances among various compartments of the body where compartments are real or fictitious spaces for drugs.

Let $x_i(t)$ be the amount of the drug in the ith compartment at time t. We shall assume that the amount that can be transferred from the ith to the jth compartment ($j \neq i$) in the time interval $(t, t + \Delta t)$ is $k_{ij}x_i(t)\Delta t + 0(\Delta t)$ where k_{ij} is called the transfer coefficent from the ith to the jth compartment. The total change Δx_i in time Δt is given by the amount entering the ith compartment from other compartments which is reduced by the amount leaving the ith compartment for other compartments including the zeroeth compartment that denotes the outside system.

Thus we get

$$\Delta x_i = - \sum_{\substack{j=0 \\ j \neq i}}^{n} k_{ij}x_i\Delta t + \sum_{\substack{j=1 \\ j \neq i}}^{n} k_{ji}x_j\Delta t + 0(\Delta t) \tag{40}$$

Dividing by Δt and proceeding to the limit as $\Delta t \to 0$, we get

$$\frac{dx_i}{dt} = -x_i \sum_{\substack{j=1 \\ j \neq i}}^{n} k_{ij} + \sum_{\substack{j=1 \\ j \neq l}}^{n} k_{ji}x_j \tag{41}$$

$$= \sum_{j=1}^{n} k_{ji}x_j, \qquad (i = 1, 2, \ldots, n, \tag{42}$$

where we define

$$k_{ii} = - \sum_{\substack{j=1 \\ j \neq i}}^{n} k_{ij}, \qquad (i = 1, 2, \ldots, n) \tag{43}$$

In matrix notation, we have

$$dX/dt = KX, \tag{44}$$

where

$$X(t) = \begin{bmatrix} x_1(t) \\ x_2(t) \\ \cdot \\ \cdot \\ \cdot \\ x_n(t) \end{bmatrix}, \quad K = \begin{bmatrix} k_{11} & k_{21} & \cdots & k_{n1} \\ k_{12} & k_{22} & \cdots & k_{n2} \\ \cdot & \cdot & \cdots & \cdot \\ k_{1n} & k_{2n} & \cdots & k_{nn} \end{bmatrix} \tag{45}$$

If $X = Be^{\lambda t}$, when B is a column matrix, (44) gives

$$\lambda Be^{\lambda t} = KBe^{\lambda t} \tag{46}$$

This gives a consistant system of equations to determine B if

$$| K - \lambda I | = 0 \tag{47}$$

where I is $n \times n$ unit matrix. Thus λ has to be an eigenvalue of the matrix K. We note that all the diagonal elements of K are negative, all the non-diagonal elements are non-negative and the sum of element of every column is greater than or equal to zero. For such a matrix, it can be shown that the real parts of the eigenvalues are always less than or equal to zero, and the imaginary part is non-zero only when the real part is strictly less than zero. Thus if $\lambda_1, \lambda_2, \ldots, \lambda_n$ are the eigenvalues then

$$Re\,(\lambda_i) \leqslant 0$$
$$Im\,(\lambda_i) \neq 0 \text{ only if } Rl\,(\lambda_i) < 0 \tag{48}$$

If the drug is injected at a constant rate given by the column vector D with components $D_1, D_2, \ldots, D_n$, (44) becomes

$$dX/dt = KX + D \tag{49}$$

Equations (44) and (49) constitute the basic equations for the analysis of drug distribution in the n-compartment system.

EXERCISE 3.3

1. Solve (44) and (49) for given initial conditions.
2. Let dose D be given at time $0, T, 2T, 3T, \ldots$, Find

$$X(nT - 0),\ X(nT + 0),\quad X(nT + t),\quad (0 < t < T)$$

3. Discuss the special cases when $n = 1, n = 2$.

3.4 MATHEMATICAL MODELLING IN ECONOMICS BASED ON SYSTEMS OF ORDINARY DIFFERENTIAL EQUATIONS OF FIRST ORDER

3.4.1 Domar Macro Model

Let $S(t)$, $I(t)$, $Y(t)$ be the Savings, Investment and National Income at time t, then it is assumed that

(i) Savings are proportional to national income, so that

$$S(t) = \alpha Y(t),\quad \alpha > 0$$

(ii) Investment is proportional to the rate of increase of national income so that

$$I(t) = \beta Y'(t),\quad \beta > 0 \tag{51}$$

(iii) All savings are invested, so that

$$S(t) = I(t) \tag{52}$$

We get a system of three ordinary differential equations of first order for determining $S(t)$, $Y(t)$, $I(t)$. Solving we get

$$Y(t) = Y(0)\,e^{\alpha t/\beta},\qquad I(t) = \alpha Y(0)e^{\alpha t/\beta} = S(t), \tag{53}$$

so that the national income, investment and savings all increase exponentially.

3.4.2 Domar First Debt Model

Let $D(t)$, $Y(t)$ denote the total national debt and total national income respectively, then we assume that

(i) Rate at which national debt changes is proportional to national income so that

$$D'(t) = \alpha Y(t) \tag{54}$$

(ii) National income increases at a constant rate, so that

$$Y'(t) = \beta \tag{55}$$

Solving

$$D(t) = D(0) + \alpha Y(0)t + \frac{1}{2}\alpha\beta t^2 \tag{56}$$

$$Y(t) = Y(0) + \beta t \tag{57}$$

so that

$$\frac{D(t)}{Y(t)} = \frac{D(0) + \alpha Y(0)t + 1/2\alpha\beta t^2}{Y(0) + \beta t} \tag{58}$$

In this model, the ratio of national debt to national income tends to increase without limit.

3.4.3 Domar's Second Debt Model

In this model, the first assumption remains the same, but the second assumption is replaced by the assumption that the rate of increase of national income is proportional to the national income so that

$$Y'(t) = \beta Y(t) \tag{59}$$

Solving (54) and (59)

$$Y(t) = Y(0)e^{\beta t} \tag{60}$$

$$D(t) = D(0) + \frac{\alpha}{\beta}Y(0)(e^{\beta t} - 1) \tag{61}$$

$$\frac{D(t)}{Y(t)} = \frac{D(0)}{Y(0)e^{\beta t}} + \frac{\alpha}{\beta}(1 - e^{-\beta t}) \tag{62}$$

In this case $D(t)/Y(t) \to \alpha/\beta$ as $t \to \infty$. Thus when debt increases at a rate proportional to income, then if the ratio of debt to income is not to increase indefinitely, income must increase exponentially.

3.4.4 Allen's Speculative Model

Let $d(t)$, $s(t)$, $p(t)$ denote the demand, supply and price of a commodity, then this model is given by

$$d(t) = \alpha_0 + \alpha_1 p(t) + \alpha_2 p'(t), \quad \alpha_0 > 0, \alpha_1 < 0, \alpha_2 > 0 \tag{63}$$

$$s(t) = \beta_0 + \beta_1 p(t) + \beta_2 p'(t), \quad \beta_0 > 0, \beta_1 > 0, \beta_2 < 0 \tag{64}$$

If $\alpha_2 = 0$, $\beta_2 = 0$ this gives Evan's price-adjustment model in which $\alpha_1 < 0$ since when price increases, demand decreases and $\beta_1 > 0$ since when price increases, supply increases. In Allen's model, coefficients α_2, β_2 account for the effect of speculation. If the price is increasing, demand increases in the expectation of the further increase in prices and supply decreases for the same reason.

For dynamic equilibrium

$$d(t) = s(t), \tag{65}$$

so that (63), (64) and (65) give

$$(\beta_2 - \alpha_2) \frac{dp}{dt} + (\beta_1 - \alpha_1) p(t) = \alpha_0 - \beta_0 \tag{66}$$

Solving

$$p(t) = p_e + (p(0) - p_e)e^{\lambda t}, \tag{67}$$

where

$$p_e = \frac{\alpha_0 - \beta_0}{\beta_1 - \alpha_1}, \quad \lambda = \frac{\alpha_1 - \beta_1}{\beta_2 - \alpha_2} \tag{68}$$

The behaviour of $p(t)$ depends on whether $p(\infty)$ or p_e is large and whether $\lambda < 0$ or $\lambda > 0$. The speculative model is highly unstable.

3.4.5 Samuelson's Investment Model

Let $K(t)$ represent the capital and $I(t)$ the investment at time t, then we assume that

(i) the investment gives the rate of increase of capital so that

$$\frac{dK}{dt} = I(t) \tag{69}$$

(ii) the deficiency of capital below a certain equilibrium level leads to an acceleration of the rate of investment proportional to this deficiency and a surplus of capital above this equilibrium level leads to a declaration of the rate of investment, again proportional to the surplus, so that

$$\frac{dI}{dt} = -m(K(t) - K_e), \tag{70}$$

where K_e is the capital equilibrium level. If $k(t) = K(t) - K_e$, we get

$$\frac{dk}{dt} = I(t), \quad \frac{dI}{dt} = -mk(t), \tag{71}$$

so that

$$-mk(t) = \frac{dI}{dt} = \frac{dI}{dk} \frac{dk}{dt} = I \frac{dI}{dk} \tag{72}$$

Integrating

$$I^2 = m(k_0^2 - k^2); \quad k_0 = k(0); \quad I(0) = 0, \tag{73}$$

so that

$$\frac{dk}{dt} = -\sqrt{m} \sqrt{k_0^2 - k^2} \tag{74}$$

and
$$k(t) = k(0) \cos \sqrt{m}\, t \tag{75}$$
$$I(t) = -k(0) \sqrt{m} \sin \sqrt{m}\, t \tag{76}$$

so that both $k(t)$ and $I(t)$ oscillate with a time period $2\pi/\sqrt{m}$.

It will be noted that if we put $k(t) = x(t)$, $I(t) = v(t)$, equations (71) are the equations for simple harmonic motion. Thus the mathematical models for the oscillation of a particle in a simple harmonic motion and for the oscillation of capital about its equilibrium value are the same.

3.4.6 Samuelson's Modified Investment Model

In this case, the rate of investment is slowed not only by excess capital as before, but it is also slowed by a high investment level so that (71) become

$$\frac{dk}{dt} = I(t), \quad \frac{dI}{dt} = -mk(t) - nI(t), \tag{77}$$

so that

$$I \frac{dI}{dk} + mk(t) + nI(t) = 0, \tag{78}$$

or

$$\frac{d^2k}{dt^2} + n\frac{dk}{dt} + mk = 0, \tag{79}$$

which are the equations for damped harmonic motion corresponding to the case when a particle performing SHM is acted as by a resistance force proportional to the velocity.

3.4.7 Stability of Market Equilibrium

Let $p_r(t)$, $s_r(t)$ and $d_r(t)$ be the price, supply and demand of a commodity in the rth market, so that Evan's price adjustment model mechanism suggests

$$\frac{dp_r}{dt} = -\mu_r(s_r - d_r), \quad r = 1, 2, \ldots, n \tag{80}$$

Now we assume that the supply and demand of the commodity in the rth market depends upon its price in all the markets, so that

$$s_r - d_r = c_r + \sum_{s=1}^{n} d_{rs}\, p_s \tag{81}$$

where c_r's and d_{rs}'s are constants. From (80) and (81), we get

$$\frac{dp_r}{dt} = -\mu_r\left(c_r + \sum_{s=1}^{n} d_{rs}p_s\right), \quad r = 1, 2, \ldots, n \tag{82}$$

If $p_{1e}, p_{2e}, \ldots p_{ne}$ are the equilibrium prices in the n markets and

$$P_r = p_r - p_{re},$$

we get

$$\frac{dP_r}{dt} = -\mu_r \sum_{s=1}^{n} d_{rs}P_s = \sum_{s=1}^{n} e_{rs}P_s, \quad r = 1, 2, \ldots, n \tag{83}$$

where $$e_{rs} = -\mu_r d_{rs} \qquad (84)$$

Substituting $P_r = A_r e^{\lambda t}$ and eliminating $A_1, A_2, \ldots, A_n$, we get

$$|\lambda I - E| = 0, \quad E = [e_{rs}] \qquad (85)$$

Thus the equilibrium will be stable if all the eigen-values of the matrix E have negative real parts.

If $d_{rs} = 0$ when $r \neq s$, the markets are independent so that non-zero value of some or all of these d_{rs}'s introduce dependence among markets.

3.4.8 Leontief's Open and Closed Dynamical Systems for Inter-industry Relations

We consider n industries. Let

x_{rs} = contribution from the rth industry to the sth industry per unit time

x_r = contribution from the rth industry to consumers per unit time

X_r = total output of the rth industry per unit time

ξ_r = input of labour in the rth industry

p_r = price per unit of the product of the rth industry

w = wage per unit of labour per unit time

Y = total labour input into the system

S_{rs} = stock of the product of the rth industry held by the sth industry

S_r = stock of the rth industry.

Thus we get the following equations:

(i) From the principle of continuity, the rate of change of stock of the rth industry = excess of the total output of the rth industry per unit time over the contribution of the rth industry to consumers and other industries per unit time, so that

$$\frac{d}{dt} S_r = X_r - x_r - \sum_{s=1}^{n} x_{rs} \qquad (86)$$

and since

$$S_r = \sum_{s=1}^{n} S_{rs} \qquad (87)$$

$$\frac{d}{dt} \sum_{s=1}^{n} S_{rs} = X_r - x_r - \sum_{s=1}^{n} x_{rs}, \quad (r = 1, 2, \ldots, n) \qquad (88)$$

(ii) Since the total labour input into the system = sum of labour inputs into all industries, we get

$$Y = \sum_{r=1}^{n} \xi_r \qquad (89)$$

(iii) Assuming the condition of perfect competition and no profit in each industry, we should have for each industry the value of input equal to the value of output so that

$$p_r X_r = \sum_{s=1}^{n} p_s x_{sr} + w \xi_r \quad (r = 1, 2, \ldots, n) \qquad (90)$$

(iv) We further assume that the input coefficients

$$a_{rs} = \frac{x_{rs}}{X_s}, \quad b_{rs} = \frac{S_{rs}}{X_s}, \quad b_r = \frac{\xi_r}{X_r} \ (r, s = 1, 2, \dots, n) \tag{91}$$

are constants.

We then get the equations

$$\frac{d}{dt} \sum_{s=1}^{n} b_{rs} X_s = X_r - x_r - \sum_{s=1}^{n} a_{rs} X_s, \quad (r = 1, 2, \dots, n) \tag{92}$$

$$Y = \sum_{s=1}^{n} b_s X_s \tag{93}$$

$$p_r = \sum_{s=1}^{n} p_s a_{sr} + w b_r, \quad (r = 1, 2, \dots, n) \tag{94}$$

We assume that the constants a_{rs}, b_{rs} b_s, are known. We also assume that $x_1, x_2, \dots, x_n$ and w are given to us as function of time, then equations (92) determine $X_1, X_2, \dots, X_n$ and then (93) determines Y and finally (94) determine $p_1, p_2, \dots, p_n$.

Thus if the final consumer's demands from all industries are known as functions of time, we can find the output which each industry must give and the total labour force required at any time. Knowing the wage rate at any time, we can find the prices of products of different industries.

EXERCISE 3.4

1. Solve Domer debt model when $Y'(t) = \beta Y^n(t)$ and deduce the two models of subsections 3.4.2 and 3.4.3 by letting $n \to 0$ and $n \to 1$. Discuss the behaviour of $D((t)/Y(t)$ as $t \to \infty$ for a general value of n.

2. Discuss the solution of Allen's speculative model when (i) $\lambda > 0$ (ii) $\lambda < 0$ (iii) $p_e > p(0)$ (iv) $p_e < p(0)$ and interpret the solution in each case.

3. Discuss the solution of Samuelson's modified investment models, when

$$\frac{dk}{dt} = I(t), \quad \frac{dI}{dt} = -mk^n(t)$$

$$\frac{dk}{dt} = I(t), \quad \frac{dI}{dt} = -mk(t) - nI^2(t)$$

4. Discuss in detail the particular case of 3.4.7 when $n = 2$.
5. Obtain the steady-state solution of Leontief's model.

3.5 MATHEMATICAL MODELS IN MEDICINE, ARMS RACE BATTLES AND INTERNATIONAL TRADE IN TERMS OF SYSTEMS OF ORDINARY DIFFERENTIAL EQUATIONS

3.5.1 A Model for Diabetes Mellitus
Let $x(t)$, $y(t)$ be the blood sugar and insulin levels in the blood stream at time t. The rate of change dy/dt of insulin level is proportional to (i) the

excess $x(t) - x_0$ of sugar in blood over its fasting level, since this exces makes the pancreas secrete insulin into the blood stream (ii) the amoun $y(t)$ of insulin since insulin left to itself tends to decay at a rate proportiona to its amount and (iii) the insulin dose $d(t)$ injected per unit time. This give

$$\frac{dy}{dt} = a_1 (x - x_0) H(x - x_0) - a_2 y + a_3 d(t), \tag{95}$$

where a_1, a_2, a_3 are positive constants and $H(x)$ is a step function which takes the value unity when $x > 0$ and taken the value zero otherwise. This occurs in (95) because if blood sugar level is less than x_0, there is no secretion of insulin from the pancreas.

Again the rate of change dx/dt of sugar level is proportional to (i) the product xy since the higher the levels of sugar and insulin, the higher is the metabolism of sugar (ii) $x_0 - x$ since if sugar level falls below fasting level, sugar is released from the level stores to raise the sugar level to normal (iii) $x - x_0$ since if $x > x_0$, there is a natural decay in sugar level proportional to its excess over fasting level (iv) function of $t - t_0$ where t_0 is the time at which food is taken

$$\frac{dx}{dt} = -b_1 xy + b_2(x_0 - x) H(x_0 - x) - b_3(x - x_0) H(x - x_0)$$

$$+ b_4 z(t - t_0), \tag{96}$$

where a suitable form for $z(t - t_0)$ can be

$$z(t - t_0) = 0, \quad t < t_0$$
$$= Q e^{-\alpha(t - t_0)}, \quad t > t_0 \tag{97}$$

Equations (95) and (96) give two simultaneous differential equations to determine $x(t)$ and $y(t)$. These equation can be numerically integrated.

3.5.2 Richardson's Model for Arms Race

Let $x(t), y(t)$ be the expenditures on arms by two countries A and B, then the rate of change dx/dt of the expenditure by the country A has a term proportional to y, since the larger the expenditure in arms by B, the larger will be the rate of expenditure on arms by A. Similarly it has a term proportional to $(-x)$ since its own arms expenditure has an inhibiting effect on the rate of expenditure on arms by A. It may also contain a term independent of the expenditures depending on mutual suspicions or mutual goodwill. With these considerations, Richardson gave the model

$$\frac{dx}{dt} = ay - mx + r, \quad \frac{dy}{dt} = bx - ny + s \tag{98}$$

Here a, b, m, n are all > 0. r and s will be positive in the case of mutual suspicions and negative in the case of mutual goodwill.

A position of equilibrium x_0, y_0, if it exists, will be given by

$$\begin{aligned} mx_0 - ay_0 - r &= 0 \\ bx_0 - ny_0 + s &= 0 \end{aligned} \quad \text{or} \quad \frac{x_0}{-as - nr} = \frac{y_0}{-br - ms}$$

$$= \frac{1}{-mn + ab}$$

or
$$x_0 = \frac{as + nr}{mn - ab}, \quad y_0 = \frac{ms + br}{mn - ab}. \tag{99}$$

If r, s are positive, a position of equilibrium exists if $ab < mn$. If $X = x - x_0$, $Y = y - y_0$, we get

$$\frac{dX}{dt} = aY - mX, \quad \frac{dY}{dt} = bX - nY \tag{100}$$

$X = Ae^{\lambda t}$, $Y = Be^{\lambda t}$ will satisfy these equations if

$$\begin{vmatrix} \lambda + m & -a \\ -b & \lambda + n \end{vmatrix} = 0, \quad \lambda^2 + \lambda(m + n) + mn - ab = 0 \tag{101}$$

Now the following cases arise:

(i) $mn - ab > 0$, $r > 0$, $s > 0$. In this case $x_0 > 0$, $y_0 > 0$ and from (101) $\lambda_1 < 0$, $\lambda_2 < 0$. As such there is a position of equilibrium and it is stable.

(ii) $mn - ab > 0$, $r < 0$, $s < 0$; there is no position of equilibrium since $x_0 < 0$, $y_0 < 0$. However since $\lambda_1 < 0$, $\lambda_2 < 0$, $X(t) \to 0$, $Y(t) \to 0$ as $t \to \infty$, so that $x(t) \to x_0$, $y(t) \to y_0$. However x_0 and y_0 are negative and populations cannot become negative. In any case to become negative, they have to pass through zero values. As such, as $x(t)$ becomes zero, (98) is modified to

$$\frac{dy}{dt} = -ny + s \tag{102}$$

and since $s < 0$, $y(t)$ decreases till it reaches zero. Similarly if $y(t)$ becomes zero first, (98) is modified to

$$\frac{dx}{dt} = -mx + r, \tag{103}$$

and since $r < 0$, $x(t)$ decreases till it reaches zero. Thus if $mn - ab > 0$, $r < 0$, $s < 0$, there will ultimately be complete disarmament.

(iii) $ma - ab < 0$, $r > 0$, $s > 0$. These give $x_0 < 0$, $y_0 < 0$, one of λ_1, λ_2 is positive and the other is negative. In this case there will be a runaway arms race.

(iv) $ma - ab < 0$, $r < 0$, $s < 0$. These give $x_0 > 0$, $y_0 > 0$ one of λ_1, λ_2 is positive and the other is negative. In this case there will be a runaway arms race or disarmament depending on the initial expenditure on arms.

3.5.3 Lanchester's Combat Model

Let $x(t)$ and $y(t)$ be the strengths of the two forces engaged in combat and let M and N be the fighting powers of individuals depending on physical fitness, types of arms and training, then Lanchester postulated that the reduction in strength of each force is proportional to the effective fighting strength of the opposite force, so that

$$\frac{dx}{dt} = -ayN, \quad \frac{dy}{dt} = -axM \tag{104}$$

giving $\qquad \dfrac{dx}{yN} = \dfrac{dy}{xM}$ or $Mx^2 - Ny^2 = \text{constant}$ $\qquad$ (105)

If the proportional reduction of strengths in the two forces are the same

$$\frac{1}{x}\frac{dx}{dt} = \frac{1}{y}\frac{dy}{dt} \quad \text{or} \quad \frac{Ny}{x} = \frac{Mx}{y} \quad \text{or} \quad Mx^2 = Ny^2 \qquad (106)$$

This is the square law. The fighting strength of an army depends on the square of its numerical strength and directly on the fighting quality of individuals.

3.5.4 International Trade Model

Since international trade is beneficial to all parties, we can consider the model

$$\frac{dx_1}{dt} = a_{12}x_1x_2 + a_{13}x_1x_3 + \ldots + a_{1n}x_1x_n$$

$$\frac{dx_2}{dt} = a_{21}x_2x_1 + a_{23}x_2x_3 + \ldots + a_{2n}x_2x_n$$

$$\cdot \quad \cdot \quad \cdot \quad \cdot \quad \cdot \quad \cdot \quad \cdot \quad \cdot \quad \cdot \qquad (107)$$

$$\frac{dx_n}{dt} = a_{n1}x_nx_1 + a_{n2}x_nx_2 + \ldots + a_{nn-1}x_nx_{n-1}$$

where all a_{ij}'s are positive. An equilibrium position is $(0, 0, \ldots, 0)$ and this is stable.

EXERCISE 3.5

1. For the Richardson's model, draw the lines $ay - mx + r = 0$, $bx - ny + s = 0$ in the four cases discussed in section 3.5.2. Draw the direction fields and possible trajectories in each case and verify the results obtained in that section.

2. For the model

$$\frac{dN_1}{dt} = N_1(a_1 - b_1N_1 - b_2N_2), \frac{dN_2}{dt} = N_2(a_2 - c_1N_1 - c_2N_n), \quad \begin{array}{l} a_1, a_2 > 0 \\ b_1, b_2 > 0 \\ c_1, c_2 > 0 \end{array}$$

find the positions of equilibrium and discuss their stability. Draw also the direction fields and possible trajectories.

3. Show that for the Lanchester model, the trajectories are hyperbolas, all of which have the same asymptotes.

4. Show that for the international trade model (107), the origin represents a position of stable equilibrium.

3.6 MATHEMATICAL MODELLING IN DYNAMICS THROUGH SYSTEMS OF ORDINARY DIFFERENTIAL EQUATIONS OF FIRST ORDER

3.6.1 Modelling in Dynamics

If a particle moves in two dimensional space, we want to determine $x(t)$,

$y(t)$, its coordinates at any time t and $u(t)$, $v(t)$ its velocity components at the same time. Similarly for the motion of a particle in three dimensions, we have to determine $x(t)$, $y(t)$, $z(t)$, $u(t)$, $v(t)$, $w(t)$. For motion of a rigid body in three dimensional space, we require twelve quantities at time t viz. six coordinates and velocities of its centre of gravity and six angles and angular velocities about the centre of gravity.

Since equation of motion are based on the principle: mass × acceleration in any direction = force in that direction, we get systems of second order differential equations. However since acceleration is the rate of change of velocity and velocity is the rate of change of displacement, we can decompose one ordinary differential equation of the second order into two ordinary differential equations of the first order.

We discuss below the motion of a particle in a plane under gravity. More general dynamical motions will be discussed in the next chapter.

3.6.2 Motion of a Projectile

A particle of mass m is projected from the origin in vacuum with velocity V inclined at an angle α to the horizontal. Suppose at time t, it is at position $x(t)$, $y(t)$ and its horizontal and vertical velocity components are $u(t)$, $v(t)$ respectively, then the equations of motion are:

$$m\frac{du}{dt} = 0 \qquad m\frac{dv}{dt} = -mg \tag{108}$$

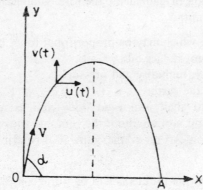

Figure 3.3

Integrating $$u = V\cos\alpha, \qquad v = V\sin\alpha - gt, \tag{109}$$

so that $$\frac{dx}{dt} = V\cos\alpha, \qquad \frac{dy}{dt} = V\sin\alpha - gt \tag{110}$$

Integrating again

$$x = V\cos\alpha t, \qquad y = V\sin\alpha t - \frac{1}{2}gt^2 \tag{111}$$

Eliminating t between these two equations, we get

$$y = x\tan\alpha - \frac{1}{2}\frac{gx^2}{V^2\cos^2\alpha} \tag{112}$$

which is a parabola, since the terms of the second degree form a perfect square. The parabola cuts $y = 0$, when

$$x = 0 \quad \text{or} \quad x = \frac{V^2 \sin 2\alpha}{g} \tag{113}$$

corresponding to position 0 and A in Figure 3.3 so that the range of the particle is given by

$$R = \frac{V^2 \sin 2\alpha}{g} \tag{114}$$

Putting $y = 0$ in (111) we get

$$t = 0 \quad \text{or} \quad t = \frac{2V \sin \alpha}{g} \tag{115}$$

This gives the time T of flight. Since the horizontal velocity is constant and equal to $V \cos \alpha$, the total horizontal distance travelled is

$$V \cos \alpha (2V \sin \alpha)/(g) = V^2 \sin 2\alpha/g$$

which gives us the same range.

3.6.3 External Ballistics of Gun Shells

To study the motion of gun shells, the following additional factors have to be taken into account:

(i) air resistance which may be proportional to v^n, but the power n can be different for different ranges of v

(ii) wind velocity, humidity and pressure

(iii) rotation of the earth

(iv) the fact that shell is a rigid body and as such both motion of its centre of gravity and motion about the centre of gravity have to be studied. When the shell comes out of the gun, it is rotating with a large angular velocity.

It is obvious that the problems will be quite complex, but all these problems have been solved and powerful computers have been developed to solve these problems because of their importance to defence.

In the case of intercontinental ballistic missiles, heating and aerodynamic effects have also to be considered.

EXERCISE 3.6

1. Show that the projectile attains the maximum height $V^2 \sin^2 \alpha/2g$ at time $V \sin/g$.

2. If the projectile is projected on a plane inclined at an angle β to the horizontal, find the range and time of flight.

3. Write the system of differential equations if there is air-resistance proportional to the nth power of the velocity. Solve the system when $n=1$.

4. Show that both the range and maximum height of a projectile are reduced by air resistance.

5. Show that with air resistance, the path of a projectile is not symmetric about the vertical line through the highest point.

6. With air resistance, which is greater:

   (i) the time of flight upto the highest point or time of flight beyond the highest point,

   (ii) the horizontal range upto the highest point or the horizontal range beyond the highest point,

and why?

4

Mathematical Modelling Through Ordinary Differential Equations of Second Order

4.1 MATHEMATICAL MODELLING OF PLANETARY MOTIONS

4.1.1 Need for the Study of Motion Under Central Forces

Every planet moves mainly under the gravitational attractive force exerted by the Sun. If S and P are masses of the Sun and the planet and G is the universal constant of gravitation, then the forces of gravitational attraction on the Sun and planet are both GSP/r^2, where r is the distance between the Sun and the planet. Accordingly the acceleration (Fig. 4.1) of the Sun towards the planet is GP/r^2 and the acceleration of the planet towards the Sun is GS/r^2. The acceleration of the planet relative to the Sun is

$$G(S + P)/r^2 = \mu/r^2.$$

Now we take the Sun as fixed, then the planet can be said to move under a central force μ/r^2 per unit mass i.e. under a force which is always directed towards a fixed centre S.

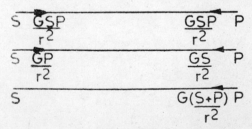

Figure 4.1

We shall for the present also regard P as a particle so that to study the motion of the planet, we have to study the motion of a particle moving under a central force. We can take S as origin so that the central force is always along the radius vector. To study this motion, it is convenient to use polar coordinates and to find the components of the velocity and acceleration along and perpendicular to the radius vector.

4.1.2 Components of Velocity and Acceleration Vectors along Radial and Transverse Directions

As the particle moves from P to Q, the displacement along the radius vector

$$= ON - OP = (r + \Delta r) \cos \Delta\theta - r \tag{1}$$

and the radial component u of velocity is

$$u = \operatorname*{Lt}_{\Delta t \to 0} \frac{(r + \Delta r) \cos \Delta\theta - r}{\Delta t}$$

$$= \operatorname*{Lt}_{\Delta t \to 0} \frac{\Delta r}{\Delta t} = \frac{dr}{dt} \tag{2}$$

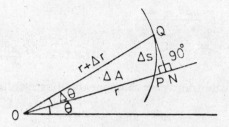

Figure 4.2

Similarly the displacement perpendicular to the radius vector

$$= (r + \Delta r) \sin \Delta\theta \tag{3}$$

and the transverse component v of the velocity is given by

$$v = \operatorname*{Lt}_{\Delta t \to 0} \frac{(r + \Delta r) \sin \Delta\theta}{\Delta t} = \operatorname*{Lt}_{\Delta t \to 0} r \frac{\sin \Delta\theta}{\Delta\theta} \frac{\Delta\theta}{\Delta t} = r \frac{d\theta}{dt} \tag{4}$$

As such the velocity components in polar coordinates are

$$u = \frac{dr}{dt} = r' \quad \text{and} \quad v = r \frac{d\theta}{dt} = r\theta' \tag{5}$$

Now the change in the velocity along the radius vector

$$= (u + \Delta u) \cos \Delta\theta - (v + \Delta v) \sin \Delta\theta - u \tag{6}$$

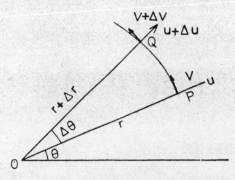

Figure 4.3

and the radial component of acceleration

$$= \underset{\Delta t \to 0}{\text{Lt}} \frac{(u + \Delta u) \cos \Delta\theta - (v + \Delta v) \sin \Delta\theta - u}{\Delta t}$$

$$= \underset{\Delta t \to 0}{\text{Lt}} \frac{\Delta u - v \Delta\theta}{\Delta t} = \frac{du}{dt} - v\frac{d\theta}{dt} = \frac{d}{dt}(r') - r\theta' \, \theta'$$

$$= r'' - r\theta'^2 \tag{7}$$

Similarly the transverse component of acceleration

$$= \underset{\Delta t \to 0}{\text{Lt}} \frac{(u + \Delta u) \sin \Delta\theta + (v + \Delta v) \cos \Delta\theta - v}{\Delta t}$$

$$= \underset{\Delta t \to 0}{\text{Lt}} \frac{u\Delta\theta + \Delta v}{\Delta t}$$

$$= u\frac{d\theta}{dt} + \frac{dv}{dt} = r'\theta' + \frac{d}{dt}(r\theta') = \frac{1}{r}\frac{d}{dt}(r^2\theta') \tag{8}$$

Thus the radial and transverse components of acceleration are

$$r'' - r\theta'^2 \quad \text{and} \quad \frac{1}{r}\frac{d}{dt}(r^2\theta') \tag{9}$$

4.1.3 Motion Under a Central Force

Let the force acting on a particle of mass m be $mF(r)$ and let it be directed towards the origin, then the equations of motion are

$$m(r'' - r\theta'^2) = -mF(r) \tag{10}$$

$$\frac{m}{r}\frac{d}{dt}(r^2\theta') = 0 \tag{11}$$

From (11)

$$r^2\theta' = \text{constant} = h \text{ (say)}, \tag{12}$$

then (10) gives

$$r'' - r\theta'^2 = -F(r) \tag{13}$$

We can eliminate t between (12) and (13) to get a differential equation between r and θ. We find it convenient to use $u = 1/r$ instead of r, so that making use of (12), we get

$$r' = \frac{dr}{dt} = \frac{dr}{du}\frac{du}{d\theta}\frac{d\theta}{dt} = -\frac{1}{u^2}\frac{du}{d\theta}\frac{h}{r^2} = -h\frac{du}{d\theta} \tag{14}$$

and

$$r'' = \frac{d}{dt}\left(-h\frac{du}{dt}\right) = \frac{d}{d\theta}\left(-h\frac{du}{d\theta}\right)\frac{d\theta}{dt}$$

$$= -h\frac{d^2u}{d\theta^2}hu^2 = -h^2u^2\frac{d^2u}{d\theta^2} \tag{15}$$

From (12), (13) and (15)

$$-F(r) = -h^2u^2\frac{d^2u}{d\theta^2} - \frac{1}{u}h^2u^4 = -h^2u^2\left(\frac{d^2u}{d\theta^2} + u\right)$$

or
$$\frac{d^2u}{d\theta^2} + u = \frac{F}{h^2u^2},\tag{16}$$

where F can be easily expressed as a function of u. This is the differential equation of the second order whose integration will give the relation between u and θ or between r and θ i.e. the equation of the path described by a particle moving under a central force F per unit mass.

4.1.4 Motion Under the Inverse Square Law

If the central force per unit mass is μ/r^2 or μu^2, Equation (16) gives

$$\frac{d^2u}{d\theta^2} + u = \frac{\mu}{h^2}\tag{17}$$

Integrating this linear equation with constant coefficients, we get

$$u = A\cos(\theta - \alpha) + \frac{\mu}{h^2}$$

or
$$\frac{h^2/u}{r} = \frac{L}{r} = 1 + e\cos(\theta - \alpha); \quad h^2 = \mu L,\tag{18}$$

which represents a conic with a focus at the centre of force. Thus if a particle moves under a central force μ/r^2 per unit mass, the path is a conic section with a focus at the centre. The conic can be an ellipse, parabola, or hyperbola according as $e \lesseqgtr 1$.

Now the velocity V of the particle is given by

$$V^2 = r'^2 + r^2\theta'^2 = \left(\frac{dr}{du}\frac{du}{d\theta}\frac{d\theta}{dt}\right)^2 + \frac{1}{u^2}(hu^2)^2$$

$$= h^2\left(\frac{du}{d\theta}\right)^2 + h^2u^2\tag{19}$$

Using (18)

$$L\frac{du}{d\theta} = -e\sin(\theta - \alpha)\tag{20}$$

From (19) and (20)

$$V^2 = \mu L\left(\frac{e^2\sin^2(\theta - \alpha)}{L^2} + \frac{(1 + e\cos(\theta - \alpha)^2}{L^2}\right)$$

$$= \frac{\mu}{L}(1 + e^2 + 2e\cos(\theta - \alpha))$$

$$= \frac{\mu}{L}(e^2 - 1 + 2(1 + e\cos(\theta - \alpha))$$

$$= \frac{\mu}{L}(e^2 - 1) + \frac{2\mu}{r}\tag{21}$$

If the path is an ellipse $\quad L = a(1 - e^2)$

If the path is a parabola $\quad e = 1$ $\tag{22}$

If the path is a hyperbola $L = a(e^2 - 1)$,

so that $V^2 = \mu \left(\dfrac{2}{r} + \dfrac{1}{a} \right)$ in the case of a hyperbola

$\qquad = \mu \left(\dfrac{2}{r} \right)$ in the case of a parabola $\hspace{2cm}$ (23)

$\qquad = \mu \left(\dfrac{2}{r} - \dfrac{1}{a} \right)$ in the case of an ellipse.

Thus if the particle is projected with velocity V from a point at a distance r from the centre of force, the path will be a hyperbola, parabola or ellipse according as

$$V^2 - \frac{2\mu}{r} \gtreqless 0 \hspace{3cm} (24)$$

We have proved that if the central force is μ/r^2 per unit mass, the path is a conic section with the centre of forces at one focus. Conversely if we know that the path is a conic section

$$\frac{L}{r} = Lu = 1 + e \cos (\theta - \alpha), \hspace{2cm} (25)$$

with a focus at the centre of force, then the force per unit mass is given by

$$F = h^2 u^2 \left(\frac{d^2 u}{d\theta^2} + u \right)$$

$$= h^2 u^2 \left(\frac{-e \cos (\theta - \alpha)}{L} + \frac{1 + \cos (\theta - \alpha)}{L} \right)$$

$$= \frac{h^2}{L} u^2 = \frac{\mu}{r^2}, \hspace{3cm} (26)$$

so that the central force follows the inverse square law.

Since all planets are observed to move in elliptic orbits with the Sun at one focus, it follows that the law of attraction between different planets and Sun must be the inverse square law.

4.1.5 Kepler's Laws of Planetory Motions

On the basis of the long period of observations of planetory motions by his predecessors and by Kepler himself, Kepler deduced the following three laws of motion empirically

(i) Every planet describes an ellipse with the Sun at one focus

(ii) The radius vector from the Sun to a planet describes equal areas in equal intervals of time.

(iii) The squares of periodic time of planets are proportional to the cubes of the semimajor axes of the orbits of the planets

We can deduce all these three laws from the mathematical modelling of planetary motion discussed above, when the law of attraction is the inverse square law.

(i) We have already seen that under the inverse square law, the path has to be a conic section and this includes elliptic orbits.

(ii) Since $r^2\theta' = h$, we get

$$\operatorname*{Lt}_{\Delta t \to 0} \frac{1}{2} \frac{r^2 \Delta\theta}{\Delta t} = \frac{1}{2}h \tag{27}$$

From Figure 4.2, the area ΔA bounded by radius vectors OP and OQ and the arc PQ is $1/2r^2 \sin \Delta\theta$ so that (27) gives

$$\frac{dA}{dt} = \frac{1}{2}h, \tag{28}$$

and the rate of description of sectorical area is constant and equal areas are described in equal intervals of time. This is Kepler's second law.

(iii) The total area of the ellipse is πab and since the areal velocity is $\frac{1}{2}h$, the periodic time T is given by

$$T = \frac{\pi ab}{\frac{1}{2}h} = \frac{2\pi ab}{\sqrt{\mu L}} = \frac{2\pi ab}{\sqrt{\mu}\sqrt{b^2/a}} = \frac{2\pi}{\sqrt{\mu}}a^{3/2} \tag{29}$$

For two different planets of masses P_1, P_2, and semiaxes of orbits a_1, a_2, this gives

$$\frac{T_1}{T_2} = \frac{\sqrt{\mu_2}}{\sqrt{\mu_1}} \frac{a_1^{3/2}}{a_2^{3/2}} = \frac{\sqrt{G(S+P_2)}}{\sqrt{G(S+P_1)}} \frac{a_1^{3/2}}{a_2^{3/2}} \tag{30}$$

or

$$\frac{T_1^2}{T_2^2} = \frac{S+P_2}{S+P_1} \frac{a_1^3}{a_2^3} = \frac{1 + \dfrac{P_2}{S}}{1 + \dfrac{P_1}{S}} \frac{a_1^3}{a_2^3} \tag{31}$$

Since P_1, P_2 are very small compared with S, this gives, as a very good approximation

$$\frac{T_1^2}{T_2^2} = \frac{a_1^3}{a_2^3} \tag{32}$$

which is Kepler's third law of planetory motion.

Deduction of Kepler's three laws of planetory motion from the universal law of gravitation was an important success of mathematical modelling. Results which took hundreds of years to obtain by observation could be obtained in a very short time by using mathematical modelling.

Here we have neglected the forces of attraction of other planets on the given planet. These are very small as compared with the attractive force of the Sun. However these can be taken into account. In fact possibly the most sensational achievement of mathematical modelling was achieved when the discrepancies from the above theory observed in the motion of planets were explained as possibly due to the existence of another small planet. The position of this planet, not observed till that time, was calculated, and when the telescope was pointed out to that position in the sky, the planet was there!

Again the occurrence of many of the fundamental particles in physics has been theoretically predicted on the basis of mathematical modelling.

The advantages of developing a successful theoretical model over relying on purely observational and empirical models are that (i) this development can suggest development of mathematical models for similar situations elsewhere and those new models can later be validated and (ii) the theoretical models, unlike empirical models, can be generalised. Thus the model developed by Newton for planetory motion could be easily extended to apply to motion of artificial satellites. Similarly in urban transportation, a gravity model was developed by trial and error and ad hoc empirical methods extending over a period of thirty to forty years. When the same model was obtained theoretically from the principle of maximum entropy, it could be easily generalised for many more complex situations than could ever be handled by the empirical methods.

EXERCISE 4.1

1. You are given the following data on orbits of major planets

Planet	Mean distance a from the Sun in millions of miles	Eccentricity e	Period T
Mercury	36.0	0.2056234	87.967 days
Venus	67.3	0.0067992	224.701 days
Earth	93.0	0.0167322	365.256 days
Mars	141.7	0.0935543	1.881 years
Jupiter	483.9	0.0484108	11.862 years
Saturn	857.1	0.0557337	29.458 years
Uranus	1785.0	0.0471703	84.015 years
Neptune	2797.0	0.0085646	164.788 years
Pluto	3670.0	0.2485200	247.697 years

(i) Show that the periods T verify Kepler's third law quite closely.

(ii) Given mass of the Sun is 2×10^{33} gms, find G

(iii) Given $G = 6.673 \times 10^{-8}$ cm³/gm sec² estimate the mass of the Sun.

(iv) Find the velocity of each planet at perihelion and apehilion.

2. Find the central force $F(r)$ if the orbit is an ellipse with the centre of force coinciding with the centre of the ellipse.

3. For a particle moving in a circular orbit of radius a, find expressions for its velocity and acceleration components.

4. Find the value of g at the surface of the Sun.

4.2 MATHEMATICAL MODELLING OF CIRCULAR MOTION AND MOTION OF SATELLITES

4.2.1 Circular Motion

When a particle moves in a circle of radius a so that $r = a$, the radial component of velocity $= r' = 0$, the transverse component of velocity $=$

$r\theta' = a\theta'$ the radial component of acceleration $= r'' - r\theta'^2 = -a\theta'^2$, the transverse component of acceleration $= \dfrac{1}{r}\dfrac{d}{dt}(r^2\theta') = \dfrac{1}{a}\dfrac{d}{dt}(a^2\theta') = a\theta''$. Thus the velocity is $a\theta'$ along the tangent and the acceleration has two components $a\theta''$ along the tangent and $a\theta'^2$ along the normal.

If a particle moves in a circle of radius a, its equations of motion are

$$ma\theta'' = \text{external force in the direction of the tangent}$$
$$ma\theta'^2 = \text{external force in the direction of the inward normal.}$$

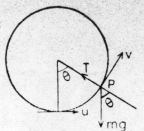

Figure 4.4

Thus if a particle is attached to one end of a string, the other end of which is fixed and the particle moves in a vertical circle, the equations of motion are (Figure 4.4)

$$ma\theta'' = -mg\sin\theta \qquad (33)$$
$$ma\theta'^2 = T - mg\cos\theta \qquad (34)$$

If θ is small, (33) gives

$$\theta'' = -\frac{g}{a}\theta, \qquad (35)$$

which is the equation for a simple harmonic motion. Thus for small oscillations of a simple pendulum, the time period is

$$T = 2\pi\sqrt{a/g} \qquad (36)$$

If θ is not necessarily small, integration of (33) gives

$$a\theta'^2 = 2g\cos\theta + \text{constant} \qquad (37)$$

If the particle is projected from the lowest point with velocity u, then $a\theta' = u$ when $\theta = 0$, so that

$$a\theta'^2 = \frac{v^2}{a} = \frac{u^2}{a} - 2g(1-\cos\theta), \qquad (38)$$

where v is the velocity of the particle, so that

$$v^2 = u^2 - 2ga(1-\cos\theta) \qquad (39)$$

or $\qquad \dfrac{1}{2}mv^2 = \dfrac{1}{2}mu^2 - mga(1-\cos\theta) = \dfrac{1}{2}mu^2 - mgh \qquad (40)$

where h is the vertical distance travelled by the particle. Equation (40) can be obtained directly from the principle of conservation of energy. Equation (34) then gives

$$T = m\frac{v^2}{a} + mg\cos\theta = m\frac{u^2}{a} - 2mg + 3\,mg\cos\theta \qquad (41)$$

At the highest point $\theta = \pi$ and $T = m\dfrac{u^2}{a} - 5mg$. If $u^2 \geqslant 5ag$, the particle will move in the complete vertical circle again and again. However if

$u^2 < 5ag$, tension will vanish before the particle reaches the highest point. When the tension vanishes, the particle begins to move freely under gravity and describes a parabolic path till the string again becomes tight and the circular motion is started again.

4.2.2. Motion of a Particle on a Smooth or Rough Vertical Wire

(a) If the particle moves on the inside of a smooth wire, the equations of motion (Fig. 4.5a) are:

$$ma\theta'' = -mg \sin \theta \tag{42}$$

$$ma\theta'^2 = R - mg \cos \theta \tag{43}$$

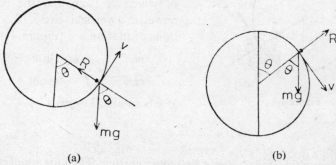

(a)   (b)

Figure 4.5

These are the same as (33) and (34) when T is replaced by the normal reaction R. As such if $u^2 \geqslant 5ag$, the particle makes an indefinite number of complete rounds of the circular wire. If $u^2 < 5ag$, the reaction vanishes before the particle reaches the highest point, the particle leaves the curve, describes a parabolic path till it meets the circular wire again and it again describes a circular path. This motion is repeated again and again.

(b) If the particle moves on the outside of the smooth vertical wire (Fig. 4.5b), the equations of motion are

$$ma\theta'' = mg \sin \theta \tag{44}$$

$$ma\theta'' = -R + mg \cos \theta \tag{45}$$

Integrating (44) $\theta'^2 = u^2 + 2ga(1 - \cos \theta)$ (46)

Using (45) $R = 3mg \cos \theta - \dfrac{mu^2}{a} - 2mg$ (47)

At the highets point $\theta = 0, R = mg - \dfrac{mu^2}{a}$ (48)

At the point A, $\theta = \pi/2, R = -\dfrac{mu^2}{a} - 2mg$ (49)

If $u^2 > ag$, the particle leaves contact with the wire immediately and describes a parabolic path.

If $u^2 < ga$, the particle remains in contact for some distance, but leaves contact when R vanishes i.e. before it reaches A and then it describes a parabolic path.

(c) If the particle moves on the inside of rough vertical circular wire, then there is an additional frictional force μR along the tangent opposing the motion. As such equations (42) and (43) are modified to

$$ma\theta'' = -mg \sin \theta - \mu R \qquad (50)$$

$$ma\theta'^2 = -mg \cos \theta + R \qquad (51)$$

Eliminating R between these equations, we get a non-linear differential equation

$$a\theta'' = -g \sin \theta - \mu(-g \cos \theta - a\theta'^2) \qquad (52)$$

which can be integrated by substituting $\theta' = w$, $\theta'' = w \, dw/d\theta$.

Similarly (44) and (45) are modified to

$$ma\theta'' = mg \sin \theta - \mu R \qquad (53)$$

$$ma\theta'^2 = -R + mg \cos \theta \qquad (54)$$

We can again eliminate R, solve for θ' and θ and find the value of θ when R vanishes.

4.2.3 Circular Motion of Satellites

Just as planets move in elliptic orbits with the Sun in one focus, the man-made artificial satellites move in elliptic (or circular) orbits with the Earth (or rather its centre) at one focus.

If the Earth is of mass M and radius a and a satellite of mass $m \, (\ll M)$ is projected from a point P at a height h above the Earth with velocity V at right angles to OP (Figure 4.6) it will move under a central force $Gm \, M/r^2$. Since the central force of a circular orbits is mV^2/r, we get, if the path is to be circular,

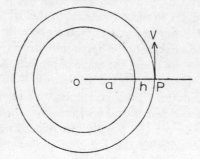

Figure 4.6

$$\frac{mV^2}{a+h} = \frac{Gm \, M}{(a+h)^2} \quad \text{or} \quad V^2 = \frac{GM}{a+h} \qquad (55)$$

If g is the acceleration due to gravity, then the gravitational force on a particle of mass m on the surface of the Earth is mg. Alternatively from Newton's inverse square law, it is GMm/a^2 so that

$$\frac{GMm}{a^2} = mg \quad \text{or} \quad Gm = ga^2 \qquad (56)$$

From (55) and (56)

$$V^2 = \frac{ga^2}{a+h} \qquad (57)$$

This gives the velocity of a satellite describing a circular orbit at a height h above the surface of the Earth. Its time period is given by

$$T = \frac{2\pi(a + h)}{V} = \frac{2\pi(a + h)}{\sqrt{ga}} (a + h)^{1/2} = \frac{2\pi}{\sqrt{ga}} (a + h)^{3/2} \qquad (58)$$

The earth completes one revolution about its axis in twenty-four hours. As such if T is 24 hours, the satellite would have the same period as the Earth and would appear stationary, to an observer on the Earth. Now taking $g = 32$ ft/sec^2, $a = 4000$ miles, $T = 24$ hours, we get if h is measured in miles

$$((4000 + h) \times 1760 \times 3)^{3/2} = \frac{24 \times 60 \times 60 \sqrt{32 \times 4000 \times 1760 \times 3 \times 7}}{2 \times 22}$$

$$= 1642607.416 \times 10^6$$

$$(4000 + h) \times 5280 = 13919.3408 \times 10^4$$

$$4000 + h = 26.36238788 \times 10^3 = 26362.38788$$

$$h = 22362.38788 \text{ miles}$$

This gives the height of the synchronous or synchron satellite, which is very useful for communication purposes.

4.2.4 Elliptic Motion of Satellites

If a satellite is projected at a height $a + h$ above the centre of the Earth with a velocity different from $\sqrt{g}\,a/\sqrt{a + h}$ or if it is not projected at right angles to the radius vector, the orbit will not be circular, but can be ellip- tic, parabolic or hyperbolic depend- ing on V and the angle of projection.

If the angle of projection is 90° and the orbit is an elliptic with semi major axis a' and eccentricity e, then there are two possibilities depending on whether the point of projection is the apogoee or the perigee Using equation (23)

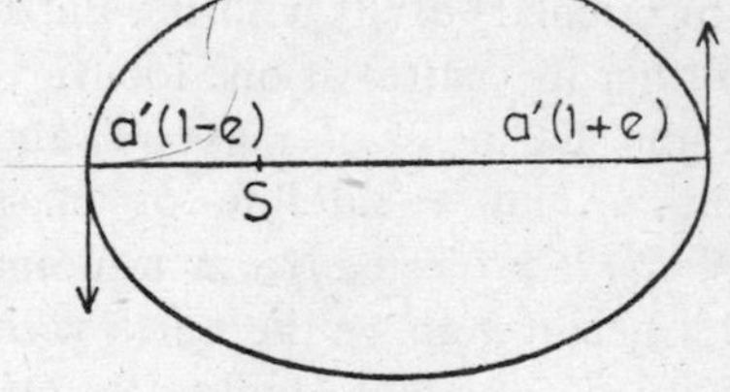

Figure 4.7

$$V^2 = \mu \left(\frac{2}{a'(1 + e)} - \frac{1}{a'} \right), \quad a'(1 + e) = a + h \qquad (59)$$

or $\qquad V^2 = \mu \left(\frac{2}{a'(1 - e)} - \frac{1}{a'} \right), \quad a'(1 - e) = a + h \qquad (60)$

i.e. $\qquad V^2 = \frac{ga^2}{a + h}(1 - e) \quad$ or $\quad V^2 = \frac{ga^2}{a + h}(1 + e)$

i.e. $\qquad V^2 = V_0^2(1 - e) \qquad$ or $\quad V^2 = V_0^2(1 + e), \qquad (61)$

where V_0 is the velocity required for a circular orbit for which $e = 0$. Thus if $V > V_0$, the point of projection is nearest point of the orbit to the centre of the Earth and if $V < V_0$, this point is the furthest point.

For the elliptic orbit, the time period is

$$T = \frac{2\pi}{\sqrt{ga}} a'^{3/2} \tag{62}$$

where if $\quad V < V_0, \; e = \sqrt{1 - \frac{V^2}{V_0^2}}, \quad a' = \frac{a + h}{1 + \sqrt{1 - V^2/V_0^2}} \tag{63}$

and if $\quad V > V_0, \; e = \sqrt{\frac{V^2}{V_0^2} - 1}, \quad a' = \frac{a + h}{1 - \sqrt{V^2/V_0^2 - 1}} \tag{64}$

If h_{max} and h_{min} are the maximum and minimum heights of a satellite above the Earth's surface and a is the radius of the Earth, we get

$$\frac{a'(1 + e)}{a'(1 - e)} = \frac{a + h_{max}}{a + h_{min}} \; \text{or} \; \frac{1 + e}{a + h_{max}} = \frac{1 - e}{a + h_{min}}$$

$$= \frac{2}{2a + h_{max} + h_{min}}$$

or $\qquad \dfrac{1 + e}{a + h_{max}} = \dfrac{1}{a + \dfrac{h_{max} + h_{min}}{2}} = \dfrac{e}{\dfrac{h_{max} - h_{min}}{2}}$

or $\qquad e = \dfrac{h_{max} - h_{min}}{2a + h_{max} - h_{min}} \tag{65}$

EXERCISE 4.2

1. Show that the force required to make a particle of mass move in a circular orbit of radius a with velocity v is mv^2/a directed towards the centre.

2. A particle of mass m is attached to the end of string, of length L, the other end of which is attached to a fixed point. The particle now moves in a horizontal circle of radius $a(< L)$. Discuss the motion of this conical pendulum.

3. Integrate (38) when $\theta' = 0$ when $\theta = \alpha$ and α is small.

4. Complete the discussion of section 4.1.1 when $u^2 = 4ag$.

5. Complete the discussion of motion of a particle on the inside of a smooth vertical circular wire when it is projected from the lowest period with horizontal velocity $2\sqrt{ag}$.

6. Complete the discussion of motion of a particle on the outside of a smooth vertical circular wire when it is projected from the highest point with velocity $3\sqrt{ag}$.

7. The following table gives data on some earth satellites

Name	max ht. (miles)	min ht. miles	weight 1bs	orbit time mts
Sputnik I	560	145	184.00	96.2
Sputnik II	1056	150	1120.00	103.7
Explorer I	1567	219	30.80	114.5

(contd.)

Vanguard	2466	405	3.25	134.0
Explorer III	1741	117	31.00	115.7
Sputnik III	1168	150	2920.00	106.0
Explorer IV	1386	178	38.43	110.0

Find the semi-major axis, semi-minor axis, eccentricity and the orbit time of each orbit and verify that the given values of the orbit times are what you expect on theoretical considerations.

8. Given $g = 981$ cm/s^2, $a = 6440 \times 10^5$ cm, $G = 6.670 \times 10^{-8}$ cm^3/(g·s^2), find the mass of the Earth.

9. Find V so that the orbit may be a parabola or a hyperbola.

4.3 MATHEMATICAL MODELLING THROUGH LINEAR DIFFERENTIAL EQUATIONS OF SECOND ORDER

4.3.1 Rectilinear Motion

Let one end 0 of an elastic string of natural length $L(= 0A)$ be fixed (Figure 4.8) and let the other end to which a particle of mass m is attached

Figure 4.8

be stretched a distance a and then released. At any time t, let $x(t)$ be the extension, then the equation of motion of the particle is

$$m \frac{d^2x}{dt^2} = -\lambda \frac{x}{L} = -kx, \qquad (66)$$

where k is the elastic constant. If the particle moves in a resisting medium with resistance proportional to the velocity x', (66) becomes

$$mx'' + cx' + kx = 0, \qquad (67)$$

which is a linear differential equation of the second order. Its solution is

$$x(t) = A_1 e^{\lambda_1 t} + A e^{\lambda_2 t} \qquad (68)$$

where λ_1, λ_2 are the roots of

$$m\lambda^2 + c\lambda + k = 0 \qquad (69)$$

Here $\lambda_1 + \lambda_2 = -\dfrac{c}{m}$, $\lambda_1\lambda_2 = \dfrac{k}{m}$. The sum of the roots is negative and the product of the roots is positive.

Case (i) $c^2 > 4km$, the roots are real and distinct and are negative. As such $x(t) \to 0$ as $t \to \infty$. The motion is said be *overdamped*.

Case (ii) $c^2 = 4$ km, the roots are real and equal and

$$x(t) = (A_1 + A_2 t) \exp\left(-\frac{c}{2m} t\right) \qquad (70)$$

and again $x(t) \to 0$ as $t \to \infty$. In this case the motion is said to be *critically damped*.

Case (iii) $c^2 < 4\,km$, the roots are complex conjugate with the real parts of the roots negative. $x(t)$ always oscillates but oscillations are damped out and tend to zero. In this case, the motion is said to be *under damped*.

Next we consider the case when there is an external force $m \cdot F(t)$ acting on the particle. In this case (67) becomes

$$mx'' + cx' + kx = mF(t) \tag{71}$$

A particular case of interest is given by the model

$$x'' + w_0^2 x = F \cos wt \tag{72}$$

i.e., when in the absence of the external force, the motion is simple harmonic with period $2\pi/w_0$ and the external force is periodic with period $2\pi/w$. The solution of (72) is given by

$$x(t) = A \cos(w_0 t - \alpha) + F \cos wt/(w_0^2 - w^2) \qquad w \neq w_0 \tag{73}$$

$$= A \cos(w_0 t - \alpha) + \frac{F}{2w_0} t \sin w_0 t \qquad w = w_0 \tag{74}$$

When $w = w_0$, the first term is periodic and its amplitude never exceeds $|A|$. However as $t \to \infty$ along a sequence for which $\sin w_0 t = \pm 1$, the magnitude of the second term approaches infinity.

The phenomenon we have discussed here is known as of *pure* or *undamped resonance*. It occurs when $c = 0$ and the input and natural frequencies are equal. We shall get a similar phenomenon when c is small. The forcing function $F \cos wt$ is then said to be in resonance with the system.

Bridges, cars, planes, ships are vibrating systems and an external periodic force with the same frequency as their natural frequency can damage them. This is the reason why soldiers crossing a bridge are not allowed to march in step. However resonance phenomenon can also be used to advantage e.g. in uprooting trees or in getting a car out of a ditch.

When w and w_0 differ only slightly, the solution represents superposition of two sinusoidal waves whose periods differ only slightly and this leads to the occurrence of beats.

4.3.2 Electrical Circuits

Figure 4.9 shows an electrical circuit. The current $i(t)$ amperes represents the time rate of change of charge q flowing in the circuits, so that

$$\frac{dq}{dt} = i(t) \tag{75}$$

(i) There is a resistance of R Ohms in the circuit. This may be provided by a light bulb, an electric heater or any other electrical device opposing the motion of the charge and causing a potential drop of magnitude $E_R = Ri$ volts.

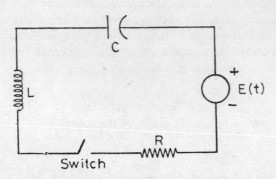

Figure 4.9

(ii) There is an induction of inductance L henrys which produces a potential drop $E_L = L \, di/dt$.

(iii) There is a capacitance C which produces a potential drop

$$E_c = \frac{1}{C} \, q.$$

All these potential drops are balanced by the battery which produces a voltage E volts. Now according to Kirchhoff's second law, the algebraic sum of the voltage drops round a closed circuit is zero so that

$$Ri + L\frac{di}{dt} + \frac{1}{C} \, q = E(t) \qquad (76)$$

Differentiating and using (75), we get

$$L\frac{d^2i}{dt^2} + R\frac{di}{dt} + \frac{1}{C} \, i = \frac{dE}{dt} \qquad (77)$$

Also substituting for (75) in (76) we get

$$L\frac{d^2q}{dt^2} + R\frac{dq}{dt} + \frac{1}{C} \, q = E(t) \qquad (78)$$

Both (77) and (78) represent linear differential equations with constant coefficients and their solutions will determine $i(t)$ and $q(t)$.

Comparing (71) and (78), we get the correspondences

mass $m \leftrightarrow$ inductance L

friction coefficient $c \leftrightarrow$ resistance R

spring constant $k \leftrightarrow$ inverse capacitance $1/C$

impressed force $F \leftrightarrow$ impressed voltage E

displacement $x \leftrightarrow$ charge q

velocity $v = dx/dt \leftrightarrow$ current $i = \dfrac{dq}{dt}.$

This shows the correspondence between mechanical and electrical systems. This forms the basis of analogue computers. A linear differential equation of the second order can be solved by forming an electrical circuit and measuring the electric current in it. Similar analogues exist between hydrodynamical and electrical systems. Mathematical modelling brings out the isomorphisms between mathematical structures of quite different systems and gives a method for solving all these models in terms of the simplest of these models.

We can have analogues of (71), (78) in economic system when $k(t)$ represents the excess of the capital invested over the equilibrium capital and $E(t)$ can represent external investments.

4.3.3 Phillip's Stabilization Model for a Closed Economy

The assumptions of the model are:

(i) The producers adjust the national production Y of a product according to the aggregate demand D. If $D > Y$, they increase production and if $D < Y$, they decrease production so that we get

$$dY/dt = \alpha(D - Y), \, \alpha > 0, \tag{79}$$

where α is a reaction coefficient representing the velocity of adjustment.

(ii) Aggregate demand D is the sum of private demand, government demand G and an exogenous disturbance u. The private demand is proportional to the national income or output so that

$$D = (1 - L) Y + G - u \tag{80}$$

where $1 - L$ is the marginal propensity to spend i.e. it is the marginal propensity to consume plus the marginal propensity to invest. We assume that $0 < L < 1$.

(iii) The government adjusts its demand to bring the national out-put to a desired level, which without loss of generality may be taken as zero.

The Government decides its demand according to one of the following policies:

(a) *proportionate stabilization policy* according to which

$$G^* = -f_p Y \tag{81}$$

where $f_p > 0$ is the coefficient of proportionality and we use the negative sign on the right hand side since if the output is less than the described level, government will come out with a positive demand.

(b) *derivative stabilization policy* according to which

$$G^* = -f_d Y', \tag{82}$$

where $f_d > 0$ and the government demand is proportional to Y'.

(c) *mixed proportionate derivative policy* according to which

$$G^* = -f_p Y - f_d Y' \tag{83}$$

(d) *integral stabilization policy* according to which

$$G^* = -f_I \int_0^t Y \, dt, \quad f_I > 0 \tag{84}$$

(iv) G^* is the potential demand which the Government may like to make, but the actual demand G will be gradually adjusted so that

$$G' = \beta(G^* - G), \tag{85}$$

where β is the reaction coefficient. $\beta > 0$ since if $G < G^*$, the government tends to increase the demand to reach G^*.

Now from (79) and (80)

$$dY/dt = \alpha((1 - L) Y + G - u - Y), \tag{86}$$

so that

$$d^2Y/dt^2 = -\alpha L \, dY/dt + \alpha \, dG/dt \tag{87}$$

Eliminating G between (85), (86) and (87)

$$\frac{d^2Y/dt^2}{\alpha} + L \, dY/dt = \beta \left(G^* - \frac{dY/dt}{\alpha} - (Ly + u) \right) \tag{88}$$

or $\qquad d^2Y/dt^2 + dY/dt \, (\alpha L + \beta) + \alpha BLY + \alpha\beta u = \alpha\beta \, G^* \tag{89}$

If we substitute for G^* from (81), (82) or (83), we get a linear differential equation of the second order with constant coefficients. If however the government uses integral stabilization policy, we use (84) to get the third order differential equation

$$d^3Y/dt^3 + (\alpha 1 + \beta) \, d^2Y/dt^2 + \alpha\beta \, dY/dt + \alpha\beta \, f_I Y = 0 \tag{90}$$

The equations (89) and (90) can be easily solved. Even without solving these, the stability of the solutions and their behaviour as $t \to \infty$ can be easily obtained.

EXERCISE 4.3

1. Solve $x'' + 13x' + 36x = 0$; $x(0) = 1$, $x'(0) = 0$ and plot $x(t)$ against t.

2. Solve $x'' + 8x' + 36x = 24 \cos 6t$ and discuss the behaviour of the solution as t approaches infinity.

3. Solve $x'' + 25x = 25 \cos 5t$ and plot $x(t)$. Discuss the nature of the motion.

4. Solve (89) for the proportionate stabilization policy. Show that the solution is

$$Y(t) = A \, e^{\lambda_1 t} + B \, e^{\lambda_2 t} + \frac{1}{1 + f_p}$$

where both λ_1, λ_2 are real and negative if $\Delta > 0$ where

$$\Delta = (\alpha L - \beta)^2 - 4\alpha\beta f_p$$

and these are complex with negative real parts of $\Delta < 0$.

5. Solve (89) for mixed proportionate-derivative stabilization policy and discuss the stability of the solution.

6. Show that all the roots of $a_0 \lambda^3 + a_1 \lambda^2 + a_2 \lambda + a_3 = 0$ have negative real parts of

$$a_1 > 0, \quad a_2 > 0, \quad a_3 > 0, \quad a_1 a_2 - a_0 a_3 > 0$$

7. Show that if (89) is solved subject to (84) and $u = 1$, the characteristic equation is

$$\lambda^3 + ((\alpha L + \beta) \lambda^2 + \alpha\beta(L + f_I)\lambda + \alpha\beta f_I = 0$$

and deduce that the stability condition is

$$f_I < (\alpha L + \beta) (L + f_p).$$

4.4 MISCELLANEOUS MATHEMATICAL MODELS THROUGH ORDINARY DIFFERENTIAL EQUATIONS OF THE SECOND ORDER

4.4.1 The Catenary

A perfectly inflexible string is suspended under gravity from two fixed points A and B (Fig. 4.10).

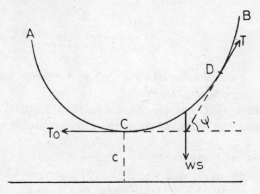

Figure 4.10

Consider the equilibrium of the part CD of the string of length s where C is the lowest point of the string at which the tangent is horizontal.

The forces acting on this part of the string are (i) tension T_0 at C (ii) tension T at point D along tangent at D (iii) weight ws of the string.

Equating the horizontal and vertical components of forces, we get

$$T \cos \psi = T_0, \qquad T \sin \psi = ws \tag{91}$$

Let T_0 be equal to weight of length c of the string, then (91) give

$$\tan \psi = \frac{ws}{T_0} = \frac{ws}{wc} = \frac{s}{c} \tag{92}$$

$$\frac{ds}{d\psi} = \rho = c \sec^2 \psi, \tag{93}$$

where ρ is radius of curvature of the string at D; so that

$$\frac{\left(1+\left(\frac{dy}{dx}\right)^2\right)^{3/2}}{\frac{d^2y}{dx^2}} = c\left(1+\left(\frac{dy}{dx}\right)^2\right)$$

or $$c\left(\frac{d^2y}{dx^2}\right) = \left(1+\left(\frac{dy}{dx}\right)^2\right)^{1/2},\tag{94}$$

which is a non-linear differential equation of second order. If $\frac{dy}{dx} = p$, then (94) gives

$$c\,\frac{dp}{\sqrt{1+p^2}} = dx\tag{95}$$

Integrating $$\sinh^{-1} p = \frac{x}{c} + A\tag{96}$$

When $x = 0, p = 0$, so that $A = 0$ and

$$\frac{dy}{dx} = \sinh\,\frac{x}{c}\tag{97}$$

Integrating

$$y = c\cosh\frac{x}{c},\tag{98}$$

where we choose x-axis in such a way that $y = c$ when $x = 0$. This is the equation of the common catenary.

It may be noted that here we get a differential equation of the second order from a problem of statics rather than from a problem of dynamics.

4.4.2 A Curve of Pursuit

A ship at the point $(a, 0)$ sights a ship at $(0, 0)$ moving along y-axis with a uniform velocity $ku(0 < k < 1)$. It begins to pursue ship B with a velocity u always moving in the direction of the ship B so that at any time AB is along the tangent to the path of A.

From Figure 4.11

$$\tan(\pi - \psi) = \frac{kut - y}{x}$$

or $$-\frac{dy}{dx} = -\frac{y}{x} + \frac{kut}{x}$$

or $$x\frac{dy}{dx} - y = -kut\tag{99}$$

Differentiating with respect to x, we get

$$x\frac{d^2y}{dx} = -ku\frac{dt}{dx}\tag{100}$$

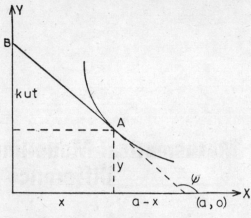

Figure 4.11

Now dx/dt = horizontal component of velocity of $A = u \cos (\pi - \psi)$

$$= -u \cos \psi = - \frac{u}{\sqrt{1 + \left(\frac{dy}{dx}\right)^2}} \qquad (101)$$

so that from (99) and (100)

$$x \frac{d^2y}{dx^2} = k \sqrt{1 + \left(\frac{dy}{dx}\right)^2} \qquad (102)$$

Putting $\frac{dy}{dx} = p$, we get

$$\frac{dp}{\sqrt{1 + p^2}} = k \frac{dx}{x} \qquad (103)$$

Integrating $\frac{dy}{dx} = k \left(\sinh^{-1} \left(\ln \frac{x}{a} \right) \right) \qquad (104)$

Integrating once again, we get y as a function of x.

EXERCISE 4.4

1. Prove that for the common catenary

$$y^2 = c^2 + s^2, \qquad\qquad s = c \tan \psi,$$

$$y = c \sec \psi \qquad\qquad s = c \sinh \frac{x}{c}$$

$$x = c \ln \frac{y + \sqrt{y^2 - c^2}}{a} = c \ln (\sec \psi + \tan \psi)$$

$$= c \ln \frac{s + \sqrt{s^2 + c^2}}{c}$$

2. Integrate (104) and find y as a function of x.
3. Obtain the curves of pursuit when $k = 1$, $k > 1$.
4. When $k < 1$, when and where does A intercept B?

Mathematical Modelling Through Difference Equations

5.1 THE NEED FOR MATHEMATICAL MODELLING THROUGH DIFFERENCE EQUATIONS: SOME SIMPLE MODELS

We need difference equation models when either the independent variable is discrete or it is mathematically convenient to treat it as a discrete variable.

Thus in Genetics, the genetic characteristics change from generation to generation and the variable representing a generation is a discrete variable.

In Economics, the price changes are considered from year to year or from month to month or from week to week or from day to day. In every case, the time variable is discretized.

In Population Dynamics, we consider the changes in population from one age-group to another and the variable representing the age-group is a discrete variable.

In finding the probability of n persons in a queue or the probability of n persons in a state or the probability of n successes in a certain number of trials, the independent variable is discrete.

For mathematical modelling through differential equations, we give an increment Δx to independent variable x, find the change Δy in y and let $\Delta x \to 0$ to get differential equations. In most cases, we cannot justify the limiting process rigorously. Thus for modelling fluid motion, making $\Delta x \to 0$ has no meaning since a fluid consists of a large number of particles and the distance between two neighbouring particles cannot be made arbitrary small. Continuum mechanics is only an approximation (through fortunately a very good one) to reality.

Even if the limiting process can be justified e.g. when the independent variable is time, the resulting differential equation may not be solvable analytically. We then solve it numerically and for this purpose, we again replace the differential equation by a system of difference equations. Numerical methods of solving differential equations essentially mean solving difference equations.

It is even argued that since in most cases, we have to ultimately solve difference equations, we may avoid modelling through differential equations altogether. This is of course going too far since as we have seen in earlier chapters, mathematical modelling through differential equations is of immense importance to science and technology. Another argument in favour of difference equation models is that those biological and social scientists who do not know calculus and transcendental numbers like e can still work with difference equation models and some important consequences of these models can be deduced with the help of even pocket calculators by even high school students.

We now give simple difference equation models parallel to the differential equation models studied in earlier chapters.

(i) *Population Growth Model*: If the population at time t is $x(t)$, then assuming that the number of births and deaths in the next unit interval of time are proportional to the populations at time t, we get the model:

$$x(t + 1) - x(t) = bx(t) - dx(t) \quad \text{or} \quad x(t + 1) = ax(t), \tag{1}$$

so that

$$x(t) = ax(t - 1) = a^2x(t - 2) = a^3x(t - 3) = \ldots = a^t x(0) \tag{2}$$

This may be compared with the differential equation model:

$$\frac{dx}{di} = ax \text{ with the solution } x(t) = x(0)e^{at} \tag{3}$$

For solving the difference equation model, we require only simple algebra, but for solving the differential equation model, we require knowledge of calculus, differential equation and exponential functions.

(ii) *Logistic Growth Model*: This is given by

$$x(t + 1) - x(t) = ax(t) - bx^2(t) \tag{4}$$

This is not easy to solve, but given $x(0)$, we can find $x(1), x(2), x(3), \ldots$ in succession and we can get a fairly good idea of the behaviour of the model with the help of a pocket calculator.

(iii) *Prey-Predator Model*: This is given by

$$\left.\begin{array}{l} x(t + 1) - x(t) = -ax(t) + bx(t)y(t) \\ y(t + 1) - y(t) = py(t) - qx(t)y(t) \end{array}\right] \begin{array}{l} a, b > 0 \\ p, q > 0 \end{array} \tag{5}$$

and again given $x(0), y(0)$, we can find $x(1), y(1); x(2), y(2); x(3), y(3), \ldots$, in succession.

(iv) *Competition Model*: This is given by

$$\left.\begin{array}{l} x(t + 1) - x(t) = ax(t) - bx(t)y(t) \\ y(t + 1) - y(t) = px(t) - qx(t)y(t) \end{array}\right] \begin{array}{l} a, b > 0 \\ p, q > 0 \end{array} \tag{6}$$

(v) *Simple Epidemics Model*: This is given by

$$\left.\begin{array}{l} x(t + 1) - x(t) = -\beta x(t)y(t) \\ y(t + 1) - y(t) = \beta x(t)y(t) \end{array}\right], \quad \beta > 0 \tag{7}$$

EXERCISE 5.1

1. For model (i), let $x(0) = 100$, $a = 0.5$ or 1 or 2; find $x(t)$ for $t = 1$ to 50 and plot $x(t)$ as a function of t in each case.

2. For model (ii) let $x(0) = 100$, $a = 0.1$, $b = 0.001$, find $x(t)$ for $t = 1$ to 100 and plot $x(t)$ as a function of t.

3. In models (iii) and (iv) let $x(0) = 40$, $y(0) = 10$, $a = 0.01$, $b = 0.001$; $p = 0.005$, $q = 0.0001$. Plot points $x(t)$, $y(t)$ for $t = 0$ to 50.

4. In model (v), let $x(0) = 100$, $y(0) = 1$, $\beta = 0.5$, plot $x(t)$, $y(t)$ in the x–y plane for $t = 0$ to 100.

5.2 BASIC THEORY OF LINEAR DIFFERENCE EQUATIONS WITH CONSTANT COEFFICIENTS

This theory is parallel to the corresponding theory of linear differential equations with constant coefficients, but is not usually taught in many places. We are therefore including a brief account here.

5.2.1 The Linear Difference Equation

An equation of the form

$$f(x_{t+n}, x_{t+n-1}, \ldots, x_t, t) = 0 \tag{8}$$

is called a difference equations of nth order. The equation

$$f_0(t)x_{t+n} + f_1(t)x_{t+n-1} + \ldots + f_n(t)x_t = \varphi(t) \tag{9}$$

is called a linear difference equation, since it involves $x_t, x_{t+1}, \ldots, x_{t+n}$ only in the first degree. The equation

$$a_0 x_{t+n} + a_1 x_{t+n-1} + \ldots + a_n x_t = \varphi(t) \tag{10}$$

is called a linear difference equation with constant coefficients. The equation

$$a_0 x_{t+n} + a_1 x_{t+n-1} + \ldots + a_n x_t = 0 \tag{11}$$

is called a homogeneous linear difference equations with constant coefficients. Let $x_t = g_1(t), g_2(t), \ldots, g_n(t)$ be n linearly independent solutions of (11), then it is easily seen that

$$x_t = A_1 g_1(t) + A_2 g_2(t) + \ldots + A_n g_n(t) \tag{12}$$

is also a solution of (11) where $A_1, A_2, \ldots, A_n$ are n arbitrary constants. This is the most general solution of (11).

Again it can be shown that if $G_1(t)$ is the solution of (11) containing n arbitrary constants and $G_2(t)$ is any particular solution of (10) containing no arbitrary constant, then $G_1(t) + G_2(t)$ is the most general solution of (10), $G_1(t)$ is called the complementary function and G_2 is called a particular solution.

5.2.2 The Complementary Function

We try the solution $x_t = a\lambda^t$. If this satisfies (11), we get

$$g(\lambda) \equiv a_0 \lambda^n + a_1 \lambda^{n-1} + a_2 \lambda^{n-2} + \ldots + a_n = 0 \tag{13}$$

This algebraic equation of nth degree has n roots $\lambda_1, \lambda_2, \ldots, \lambda_n$, real or complex. The complementary function is then given by

$$G_1(t) = c_1\lambda_1^t + c_2\lambda_2^t + \ldots + c_n\lambda_n^t \tag{14}$$

Case (i):   If $\lambda_1, \lambda_2, \ldots, \lambda_n$ are all real and distinct, (14) gives us the complementary function when $c_1, c_2, \ldots, c_n$ are any n arbitrary real constants.

Case (ii):   If two of the roots λ_1, λ_2 are equal, then (14) contains only $n - 1$ arbitrary constants and as such it cannot be the most general solution. We try the solution $ct\lambda_1^t$. We get

$$a_0(t + n)\lambda_1^n + a_1(t + n - 1)\lambda_1^{n-1} + \ldots + a_n = 0$$

or $\qquad tg(\lambda_1) + g'(\lambda_1) = 0,$ $\hfill (15)$

which is identically satisfied since both $g(\lambda_1) = 0$ and $g'(\lambda_1) = 0$ as λ_1 is a repeated root. In this case

$$G_1(t) = (c_1 + c_2t)\lambda_1^t + c_3\lambda_3^t + c_4\lambda_4^t + \ldots + c_n\lambda_n^t \tag{16}$$

Case (iii):   If a root λ_1 is repeated k times, the complementary function is

$$G_1(t) = (c_1 + c_2t + c_3t^2 + \ldots + c_kt^{k-1})\lambda_1^t + c_{k+1}\lambda_{k+1}^t$$
$$+ \ldots + c_n\lambda_n^t \tag{17}$$

Case (iv):   Let $g(\lambda) = 0$ have two complex roots $\alpha \pm i\beta$, then their contribution to complementary function is

$$c_1(\alpha + i\beta)^t + c_2(\alpha - i\beta)^t \tag{18}$$

Putting $\alpha = r\cos\theta, \beta = r\sin\theta$ and using De Moivre's theorem, this reduces to

$$c_1r^t(\cos\theta + i\sin\theta)^t + c_2r^t(\cos\theta - i\sin\theta)^t$$
$$= r^t\cos(\theta t)(c_1 + c_2) + r^t\sin(\theta t)(ic_1 - ic_2)$$
$$= r^t(d_1\cos(\theta t) + d_2\sin(\theta t))$$
$$= (\alpha^2 + \beta^2)^{t/2}(d_1\cos(\theta t) + d_2\sin(\theta t)), \tag{19}$$

where $\tan\theta = \dfrac{\beta}{\alpha}$ $\hfill (20)$

and d_1, d_2 are arbitrary constants.

Case (v):   If the complex roots $\alpha \pm i\beta$ are repeated k times, then contribution to the complementary function is

$$(\alpha^2 + \beta^2)^{t/2}((d_0 + d_1t + \ldots + d_{k-1}t^{k-1}\cos(\theta t)$$
$$+ (f_0 + f_1t + \ldots + f_{k-1}t^{k-1})\sin(\theta t) \tag{21}$$

where $d_0, d_1, \ldots, d_{k-1}, f_0, \ldots, f_{k-1}$ are $2k$ arbitrary constants.

5.2.3 The Particular Solution
Here we want a solution of (10) not containing any arbitrary constant.

Case (i): $\qquad$ Let $\varphi(t) = AB^t$, B is not a root of $g(\lambda) = 0$ $\hfill (22)$

We try the solution CB^t. Substituting in (10), we get

$$CB^t(a_0B^n + a_1B^{n-1} + \ldots + a_n) = AB^t \tag{23}$$

If $B \neq \lambda_1, \lambda_2, \ldots, \lambda_n$, we get

$$C = \frac{A}{a_0B^n + a_1B^{n-1} + \ldots + a_n} \tag{24}$$

and the particular solution is

$$\frac{AB^t}{a_0B^n + a_1B^{n-1} + \ldots + a_n} \tag{25}$$

Case (ii):   Let

$$\varphi(t) = AB^t, B \text{ is a non-repeated root of } g(\lambda) = 0 \tag{26}$$

We try the solution CtB^t. Substituting in (10), we get

$$B^t(Ct\, g(B) + Cg'(B)) = AB^t \tag{27}$$

Since $g(B) = 0, g'(B) \neq 0$

$$C = \frac{A}{g'(B)}, \tag{28}$$

so that the particular solution is

$$\frac{AtB^t}{a_0nB^{n-1} + a_1(n-1)B^{n-2} + \ldots + a_{n-1}} \tag{29}$$

Case (iii):   Let

$$\varphi(t) = AB^t, \quad g(B) = 0, \quad g'(B) = 0, \ldots,$$
$$g^{(k-1)}(B) = 0, \qquad g^{(k)}(B) \neq 0, \tag{30}$$

then the particular solution is

$$\frac{At^{k-1}B^t}{g^{(k)}(B)} \tag{31}$$

Case (iv):   Let $\qquad\qquad \varphi(t) = At^k \tag{32}$

We try the solution

$$d_0t^k + d_1t^{k-1} + d_2t^{k-2} + \ldots + d_k \tag{33}$$

Substituting in (10) we get

$$a_0(d_0(t+n)^k + d_1(t+n)^{k-1} + d_2(t+n)^{k-2} + \ldots + d_k)$$
$$+ a_1(d_0(t+n-1) + d_1(t+n-1)^{k-1} + d_2(t+n-1)^{k-2}$$
$$+ \ldots + d_k) + \ldots + a_n(d_0t^k + d_1t^{k-1} + d_2t^{k-2} + \ldots + d_k)$$
$$= 0 \tag{34}$$

Equating the coefficients of t^k, t^{k-1}, $\ldots$, t^0, on both sides, we get $(k+1)$ equations which in general will enable us to determine $d_0, d_1, d_2, \ldots, d_k$ and thus the particular solution will be determined.

5.2.4 Obtaining Complementary Function by Use of Matrices

Let

$$x_t = x_1(t)$$
$$x_{t+1} = x_2(t) = x_1(t + 1)$$
$$x_{t+2} = x_3(t) = x_2(t + 1) \tag{35}$$
$$\cdots \quad\quad \cdots \quad\quad \cdots$$
$$x_{t+n} = x_{n+1}(t) = x_n(t + 1),$$

so that (11) becomes

$$a_0 x_n(t + 1) = -a_1 x_n(t) - a_2 x_{n-1}(t) - \ldots - a_n x_1(t) \tag{36}$$

Equations (35) and (36) give

$$x_1(t + 1) = x_2(t)$$
$$x_2(t + 1) = x_3(t)$$
$$\cdots \quad\quad \cdots \tag{37}$$
$$x_{n-1}(t + 1) = x_n(t)$$
$$x_n(t + 1) = -\frac{a_1}{a_0} x_n(t) - \frac{a_2}{a_0} x_{n-1}(t) - \ldots - \frac{a_n}{a_0} x_1(t),$$

which can be written in the matrix form

$$
\begin{bmatrix}
x_1(t + 1) \\
x_2(t + 1) \\
\cdot \\
\cdot \\
\cdot \\
x_n(t + 1)
\end{bmatrix}
$$

$$
\begin{bmatrix}
0 & 1 & 0 & \cdots & 0 \\
0 & 0 & 1 & \cdots & 0 \\
\cdot\cdot & \cdot\cdot & \cdot\cdot & \cdots & \cdot\cdot \\
-\dfrac{a_n}{a_0} & -\dfrac{a_{n-1}}{a_0} & -\dfrac{a_{n-2}}{a_0} & \cdots & -\dfrac{a_1}{a_0}
\end{bmatrix}
\begin{bmatrix}
x_1(t) \\
x_2(t) \\
\cdot \\
\cdot \\
x_n(t)
\end{bmatrix}
\tag{38}
$$

or

$$X(t + 1) = AX(t), \tag{39}$$

where

$$
X(t) =
\begin{bmatrix}
x_1(t) \\
x_2(t) \\
\cdot \\
\cdot \\
\cdot \\
x_n(t)
\end{bmatrix},
$$

$$
A =
\begin{bmatrix}
0 & 1 & 0 & \cdots & 0 \\
0 & 0 & 1 & \cdots & 0 \\
\cdot\cdot & \cdot\cdot & \cdot\cdot & \cdots & \cdot\cdot \\
0 & 0 & 0 & \cdots & 1 \\
-\dfrac{a_n}{a_0} & -\dfrac{a_{n-1}}{a_0} & -\dfrac{a_{n-2}}{a_0} & \cdots & -\dfrac{a_1}{a_0}
\end{bmatrix}
\tag{40}
$$

Applying (39) repeatedly

$$X(k) = A^k X(0),$$ (41)

where

$$X(0) = \begin{bmatrix} x_1(0) \\ x_2(0) \\ x_3(0) \\ \cdot \\ \cdot \\ \cdot \\ x_n(0) \end{bmatrix} = \begin{bmatrix} x_1(0) \\ x_1(1) \\ x_1(2) \\ \cdot \\ \cdot \\ \cdot \\ x_1(n-1) \end{bmatrix} = \begin{bmatrix} x_0 \\ x_1 \\ x_2 \\ \cdot \\ \cdot \\ \cdot \\ x_{n-1} \end{bmatrix}$$ (42)

Thus knowing the values of x_1 at times $0, 1, 2. \ldots, n-1$, we can find its value at all subsequent times.

5.2.5 Solution of a System of Linear Homogeneous Difference Equations with Constant Coefficients

Let the system be given by

$$x_1(t + 1) = a_{11}x_1(t) + a_{12}x_2(t) + \ldots + a_{1n}x_n(t)$$
$$x_2(t + 1) = a_{21}x_1(t) + a_{22}x_2(t) + \ldots + a_{2n}x_n(t)$$
$$\cdots \qquad \cdots \qquad \cdots \qquad \cdots \qquad \cdots$$
$$x_n(t + 1) = a_{n1}x_1(t) + a_{n2}x_2(t) + \ldots + a_{nn}x_n(t)$$ (43)

This can be written in the matrix form

$$X(t + 1) = AX(t),$$ (44)

where

$$X(t) = \begin{bmatrix} x_1(t) \\ x_2(t) \\ \cdot \\ \cdot \\ x_n(t) \end{bmatrix}, \quad A = \begin{bmatrix} a_{11} & a_{12} & \ldots & a_{1n} \\ a_{21} & a_{22} & \ldots & a_{2n} \\ \cdot & \cdot & \cdot & \cdot \\ a_{n1} & a_{n2} & \ldots & a_{nn} \end{bmatrix}$$ (45)

Applying (44) repeatedly, we wet

$$X(k) = A^k X(0)$$ (46)

5.2.6 Solution of Linear Difference Equations by Using Laplace Transform

Let the linear difference equation be

$$a_0 f(t) + a_1 f(t - 1) + \ldots + a_n f(t - n) = \varphi(t),$$
$$f(t) = 0 \quad \text{when } t < 0$$ (47)

Let $\bar{f}(\lambda)$ be the Laplace transform of $f(t)$ so that

$$\bar{f}(\lambda) = L(f(t)) = \int_0^\infty e^{-\lambda t} f(t)\, dt$$ (48)

then $L(f(t-1)) = \int_1^\infty e^{-\lambda t} f(t-1)\, dt$

$$= e^{-\lambda} \int_0^\infty e^{-\lambda t} f(t)\, dt = e^{-\lambda} \bar{f}(\lambda)$$

$$L(f(t-2)) = \int_2^\infty e^{-\lambda t} f(t-2)\, dt$$

$$= e^{-2\lambda} \int_0^\infty e^{-\lambda t} f(t)\, dt = e^{-2\lambda} \bar{f}(\lambda) \tag{49}$$

and so on so that taking Laplace transform of both sides of (49), we get

$$(a_0 + a_1 e^{-\lambda} + a_2 e^{-2\lambda} + \ldots + a_n e^{-n\lambda})\bar{f}(\lambda) = L(\varphi(t)) = \bar{\varphi}(\lambda), \tag{50}$$

so that $\bar{f}(\lambda)$ is known. Inverting the Laplace transform, we get $f(t)$. In this case t is regarded as a continuous variate such that $f(t) = 0$ when $t < 0$. If t is a discrete variate, it is better to use the z-transform.

5.2.7 Solution of Linear Difference Equations by Using z-Transform

Let $\{u_n\}$ be an infinite sequence, then its z-transform is defined by

$$Z(u_n) = \sum_{n=0}^\infty u_n z^{-n}, \tag{51}$$

whenever this infinite series converges. If $\{u_n\}$ is a probability distribution and $z = 1/s$, it will be the same as the probability generating function. The following results can be easily established

(i) If $k > 0$, $Z(u_{n-k}) = z^{-k} Z(u_n)$ $\tag{52}$

(ii) If $k > 0$, $Z(u_{n+k}) = z^k[Z(u_n) - \sum_{m=0}^{k-1} u_m z^{-m}]$ $\tag{53}$

(iii) u_n : $\qquad 1 \qquad\qquad a^n \qquad\qquad e^{an}$

$\quad Z(u_n)$: $\quad z/(z-1) \quad z/(z-a) \quad z/(z-e^a)$ $\tag{54}$

Taking z-transform of both sides of a linear difference equation, we can find $Z(u_n)$ and expanding it in powers of $1/z$ and finding the coefficient of z^{-n}, we can get u_n.

5.2.8 Solution of non-Linear Difference Equations Reducible to Linear Equations

Thus equations

$$y_{n+1} = \sqrt{y_n} \tag{55}$$

$$y_n y_{n+2} = y_{n+1}^2 \tag{56}$$

become linear on substitution $u_n = \ln y_n$. Also

$$y_{n+2} = \frac{y_n y_{n+1}}{y_n + y_{n+1}} \tag{57}$$

becomes linear on substitution $u_n = 1/y_n$.

5.2.9 Stability Theory for Difference Equations

If $x_t = K$ satisfies

$$f(x_t, x_{t+1}, x_{t+2}, \ldots, x_{t+n}) = 0 \tag{58}$$

then this gives an equilibrium position. To find its stability, we substitute $x_t = K + u_t$ in (58) and simplify neglecting squares and products and higher powers of u_t's to get a linear equation

$$a_1 u_{t+n} + a_2 u_{t+n-1} + \ldots + a_n u_t = 0 \tag{59}$$

We try the solution $u_t = A\lambda^t$ and get the characteristic equation

$$a_0 \lambda^n + a_1 \lambda^{n-1} + \ldots + a_n = 0 \tag{60}$$

If the absolute value of each of the n roots of this equation is less than unity, then u_t would tend to zero as $t \to \infty$ for all small initial disturbances and the equilibrium position would be locally asymptotically stable.

The conditions for all the roots of (60) having magnitude less than unity are given by Schur's criterion viz. that all the following determinants should be positive.

$$\Delta_1 = \begin{vmatrix} a_0 & a_n \\ a_n & a_0 \end{vmatrix}, \qquad \Delta_2 = \begin{vmatrix} a_0 & 0 & \cdot & a_n & a_{n-1} \\ a_1 & a_0 & \cdot & 0 & a_n \\ \cdot & \cdot & \cdot & \cdot & \cdot \\ a_n & 0 & \cdot & a_0 & a_1 \\ & & \cdot & & \\ a_{n-1} & a_n & \cdot & 0 & a_0 \end{vmatrix}$$

$$\Delta_n = \begin{vmatrix} a_0 & 0 & \cdots & 0 & \cdot & a_n & a_{n-1} & \cdots & a_1 \\ & & & & \cdot & & & & \\ a_1 & a_0 & \cdots & 0 & \cdot & 0 & a_n & \cdots & a_2 \\ \cdot\cdot & \cdot\cdot & \cdots & \cdot\cdot & \cdot & \cdot\cdot & \cdot\cdot & \cdots & \cdot\cdot \\ a_{n-1} & a_n & \cdots & a_0 & \cdot & 0 & 0 & \cdots & a_n \\ \cdot\cdot & \cdots\cdots & \cdots & \cdots & \cdots & \cdots & \cdots & \cdots & \cdot\cdot \\ a_n & 0 & \cdots & 0 & \cdot & a_0 & a_1 & \cdots & a_{n-1} \\ & & & & \cdot & & & & \\ a_n & a_n & \cdots & 0 & \cdot & 0 & a_0 & \cdots & a_{n-2} \\ \cdot\cdot & \cdot\cdot & \cdots & \cdot\cdot & \cdot & \cdot\cdot & \cdot\cdot & \cdots & \cdot\cdot \\ a_1 & a_2 & \cdots & a_n & \cdot & 0 & 0 & \cdots & a_0 \end{vmatrix}$$

$$\tag{61}$$

EXERCISE 5.2

1. Solve the following and discuss the behaviour of each solution as $t \to \infty$:

(i) $x_{t+2} - 7x_{t+1} + 12x_t = 0$

(ii) $x_{t+3} - 5x_{t+2} + 7x_{t+1} - 3x_t = 0$

(iii) $x_{t+2} - 2x_{t+1} + 2x_t = 0$

(iv) $8x_{t+3} - 12x_{t+2} + 6x_{t+1} - x_t = 0$

(v) $x_{t+2} + 2x_{t+1} + x_t = 0$

(vi) $2x_{t+2} - 2x_{t+1} + x_t = 0$

(vii) $x_{t+2} - x_{t+1} + x_t = 0$

2. Solve the following difference equations

(i) $x_{t+2} - 4x_{t+1} + 4x_t = 2^t$

(ii) $x_{t+2} - 4x_{t+1} + 3x_t = t$

(iii) $x_{t+2} - 7x_{t+1} + 12x_t = 3^t + t^4 + 4^t t^3$.

3. Solve the following simultaneous equations

(i) $x_{n+1} - x_n + 2y_{n+1} = 0$

$y_{n+1} - y_n - 2x_n = 2^n$

(ii) $x_{n+1} - 2x_n - y_n = n$

$y_{n+1} - 2x_n - 3y_n = -n$

4. Solve difference equations in Exercises 1 and 2 by using

(i) Laplace transform method

(ii) Z-transform method

(iii) Transforming to a matrix equation.

5. Prove results (52), (53), (54) and solve equations (55), (56), (57).

6. Show that the system (44) will be stable if all the eigenvalue of this matrix have magnitude less than unity.

7. Prove that for (44) to be stable, it is necessary that

$$| A | < 1, \; -n < \text{trace } A < n$$

8. Prove that if the sum of the elements of each column of a square matrix with non-negative elements is less than unity, then all the characteristic roots of this matrix have magnitude less than unity.

9. Discuss the stability of the following systems

(i) $x_{t+3} + 9x_{t+2} - 5x_{t+1} - 2x_t = 0$

(ii) $2x_{t+2} - 2x_{t+1} + x_t = 0$

(iii) $\begin{bmatrix} x_{t+1} \\ y_{t+1} \\ z_{t+1} \end{bmatrix} = \begin{bmatrix} 6 & -11 & 6 \\ 1 & 0 & 0 \\ 0 & 1 & 0 \end{bmatrix} \begin{bmatrix} x_t \\ y_t \\ z_t \end{bmatrix}$

10. Write explicitly the conditions that all roots of

(i) $a_0\lambda^2 + a_1\lambda + a_2 = 0$ (ii) $a_0\lambda^3 + a_1\lambda^2 + a_2\lambda + a_3 = 0$

are less than unity in magnitude.

5.3 MATHEMATICAL MODELLING THROUGH DIFFERENCE EQUATIONS IN ECONOMICS AND FINANCE

6.3.1 The Harrod Model

Let $S(t)$, $Y(t)$, $I(t)$ denote the savings, national income and investment respectively. We make now the following assumptions:

(i) Savings made by the people in a country depend on the national income i.e.

$$S(t) = \alpha Y(t), \quad \alpha > 0 \tag{62}$$

(ii) The investment depends on the difference between the income of the current year and the last year i.e.

$$I(t) = \beta(Y(t) - Y(t - 1)), \qquad \beta > 0 \tag{63}$$

(iii) All the savings made are invested, so that

$$S(t) = I(t) \tag{64}$$

From (62), (63) and (64), we get the difference equation

$$Y(t) = \frac{\beta}{\beta - \alpha} Y(t - 1), \tag{65}$$

which has the solution

$$Y(t) = A\left(\frac{\beta}{\beta - \alpha}\right)^t = Y(0)\left(\frac{\beta}{\beta - \alpha}\right)^t \tag{66}$$

Assuming that $Y(t)$ is always positive,

$$\beta > \alpha, \beta/(\beta - \alpha) > 1, \tag{67}$$

so that the national income increases with t. The national incomes at different times 0, 1, 2, 3, ... form a geometrical progression.

Thus if all savings are invested, savings are proportional to national income and the investment is proportional to the excess of the current years income over the preceding years income, then the national income increases geometrically.

5.3.2 The Cobweb Model

Let p_t = price of a commodity in the year t and

q_t = amount of the commodity available in the market in year t, then we make the following assumptions

(i) Amount of the commodity produced this year and available for sale is a linear function of the price of the commodity in the last year, i.e.

$$q_t = \alpha + \beta p_{t-1}, \tag{68}$$

where $\beta > 0$ since if the last year's price was high, the amount available this year would also be high.

(ii) The price of the commodity this year is a linear function of the amount available this year i.e.

$$p_t = \gamma + \delta q_t, \tag{69}$$

where $\delta < 0$, since if q_t is large, the price would be low. From (68) and (69)

$$p_t - \beta\delta\, p_{t-1} = \gamma + \alpha\delta, \tag{70}$$

which has the solution

$$\left(p_t - \frac{\alpha\delta + \gamma}{1 - \beta\delta}\right) = \left(p_0 - \frac{\alpha\delta + \gamma}{1 - \beta\delta}\right)(\beta\delta)^t, \tag{71}$$

so that

$$\left(p_t - \frac{\alpha\delta + \gamma}{1 - \beta\delta}\right) = \left(p_{t-1} - \frac{\alpha\delta + \gamma}{1 - \beta\delta}\right)(\beta\delta) \tag{72}$$

Since $\beta\delta$ is negative $p_0, p_1, p_2, p_3, \ldots$ are alternatively greater and less than $(\alpha\delta + \gamma)/(1 - \beta\delta)$.

If $|\beta\delta| > 1$, the deviation of p_t from $(\alpha\delta + \gamma)/(1 - \beta\delta)$ goes on increasing. On the other hand if $|\beta\delta| < 1$, this deviation goes on decreasing and ultimately $p_t \to (\alpha\delta + \gamma)/(1 - \beta\delta)$ as $t \to \infty$.

Figures 5.1a and 5.1b show how the price approaches the equilibrium price $p_e = (\alpha\delta + \gamma)/(1 - \beta\delta)$ as t increases in the two cases when $p_0 > p_e$ and $p_0 < p_e$ respectively.

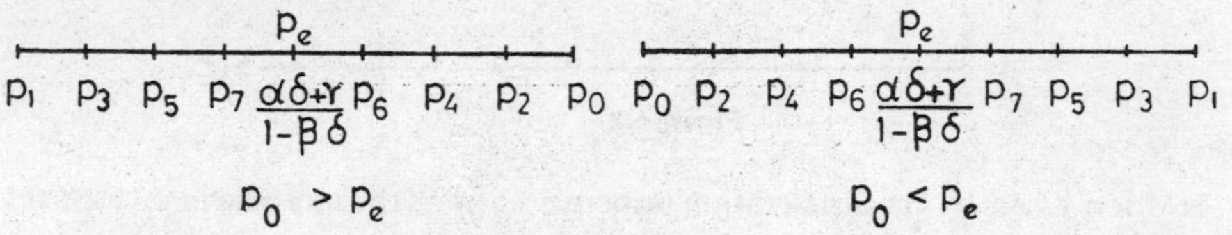

Figure 5.1

In the same way, eliminating p_t from (67), (68) we get

$$q_t = \alpha + \beta\gamma + \beta\delta\, q_{t-1}, \tag{73}$$

which has the solution

$$\left(q_t - \frac{\alpha + \beta\gamma}{1 - \beta\delta}\right) = \left(q_e - \frac{\alpha + \beta\gamma}{1 - \beta\delta}\right)(\beta\delta)^t, \tag{74}$$

so that q_t also oscillates about the equilibrium quantity level

$$q_t = (a + \beta\gamma)/(1 - \beta\delta) \quad \text{if} \quad |\beta\delta| < 1$$

The variation of both prices and quantities is shown simultaneously in Figure 5.2.

Suppose we start in the year zero with price p_0, and quantity q_0 represented by the point A. In year 1, the quantity q_1 is given by $\alpha + \beta p_0$ and the price is given by $p_1 = \gamma + \delta q_1$. This brings us to the point C in two steps via B. The path of prices and quantities is thus given by the Cobweb path $ABCDEFGHI, \ldots$ and the equilibrium price and quantity are given by the intersection of the two straight lines.

5.3.3 Samuelson's Interaction Models

The basic equations for the first interaction model are:

$$Y(t) = C(t) + I(t), \quad C(t) = \alpha Y(t - 1), \quad I(t) = \beta[C(t) - C(t - 1)] \tag{75}$$

Here the positive constant α is the marginal propensity to consume with respect to income of the previous year and the positive constant β is the

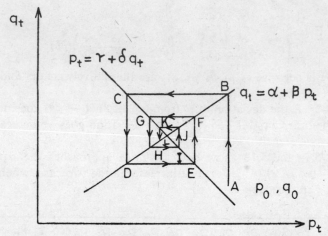

Figure 5.2

relation given by the acceleration principle i.e. β is the increase in investment per unit of excess of this years consumption over the last year's.

From (75), we get the second order difference equation

$$Y(t) - \alpha(1 + \beta)Y(t - 1) + \alpha\beta Y(t - 2) = 0 \qquad (76)$$

In the second interaction model, there is an additional investment by the government and this investment is assumed to be a constant γ. In this case (76) is modified to

$$Y(t) - \alpha(1 + \beta)Y(t - 1) + \alpha\beta Y(t - 2) - \gamma = 0 \qquad (77)$$

The solution of (76) and (77) can show either an increasing trend in $Y(t)$ or a decreasing trend in $Y(t)$ or an oscillating trend in it.

5.3.4 Application to Actuarial Science

One important aspect of actuarial science is what is called mathematics of finance or mathematics of investment.

If a sum S_0 is invested at compound interest of i per unit amount per unit time and S_t is the amount at the end of time t, then we get the difference equation

$$S_{t+1} = S_t + iS_t = (1 + i)S_t, \qquad (78)$$

which has the solution

$$S_t = S_0(1 + i)^t, \qquad (79)$$

which is the well-known formula for compound interest.

Suppose a person borrows a sum S_0 at compound interest i and wants to amortize his debt, i.e. he wants to pay the amount and interest back by payment of n equal instalments, say R, the first payment to be made at the end of the first year.

Let S_t be the amount due at the end of t years, then we have the difference equation

$$S_{t+1} = S_t + iS_t - R = (1 + i)S_t - R \tag{80}$$

Its solution is

$$S_t = \left(S_0 - \frac{R}{i}\right)(1 + i)^t + \frac{R}{i} \tag{81}$$

$$= S_0(1 + t)^t - R\,\frac{(1 + i)^t - 1}{i} \tag{82}$$

If the amount is paid back in n years, $S_n = 0$, so that

$$R = S_0\frac{i}{1 - (1 + i)^{-n}} = S_0\,\frac{1}{a_{\overline{n}|i}}, \tag{83}$$

where $a_{\overline{n}|i}$ called the amortization factor is the present value of an annuity of 1 per unit time for n periods at an interest rate i.

The functions $a_{\overline{n}|i}$ and $(a_{\overline{n}|i})^{-1}$ are tabulated for common values of n and i.

Suppose an amount R is deposited at the end of every period in a bank and let S_t be the amount at the end of t periods, then

$$S_{t+1} = S_t(1 + i) + R, \tag{84}$$

so that (since $S_0 = 0$)

$$S_n = R\,\frac{(1 + i)^n - 1}{i} = RS_{\overline{n}|i} \tag{85}$$

From (83) and (85)

$$S_{\overline{n}|i} = (1 + i)^n a_{\overline{n}|i} \tag{86}$$

or

$$\frac{1}{S_{\overline{n}|i}} = \frac{(1 + i)^{-n}}{a_{\overline{n}|i}} \tag{87}$$

If a person has to pay an amount S at the end of n years, he can do it by paying into a sinking find an amount R per period where

$$R = S\,\frac{1}{S_{\overline{n}|i}} \tag{88}$$

where $\dfrac{1}{S_{\overline{n}|i}}$ is the sinking fund factor and can be tabulated by using (87).

EXERCISE 5.3

1. Show that the necessary and sufficient conditions for both roots of

$$m^2 + a_1m + a_2 = 0$$

to be less than unity in absolute magnitude are

$$1 + a_1 + a_2 > 0, \quad 1 - a_1 + a_2 > 0, \quad 1 - a_2 > 0$$

2. Use the condition of Ex. 1 to show that the model of equation (76) is stable if

$$1 - \alpha > 0, \quad 1 - \alpha\beta > 0$$

i.e. if both the marginal propensity to consume and its product with the relation must be less than unity.

3. Show that if the condition of Ex. 2 are satisfied, then for the model of equation (77), the national income will tend to its equilibrium value $\gamma/(1 - \alpha)$. Show also that the approach to equilibrium value will be oscillatory if

$$\alpha(1 + \beta)^2 < 4\alpha\beta$$

4. For the model

$$Y_t = I_t + C_t, \quad C_t = C + mY_t, \quad rI_t = Y_{t+1} - Y_t$$

find C_t, I_t, Y_t and discuss stability of equilibrium position.

5. Let S_t denote the amount due at the end of t periods when the amounts being paid are $R, 2R, 3R, \ldots$. Show that

$$S_{t+1} = S_t(1 + i) + (t + 1)R$$

Show that the solution is

$$S_t = \frac{R}{i} [(1 + i)S_{\overline{t}|i} - t]$$

6. Discuss the extended Cohweb model for which

$$p_t - p_e = c(1 - P)(p_{t-1} - p_e) + cP(p_{t-2} - p_e),$$

where c is the ratio of slopes of supply and demand curves and P (usually $0 \leqslant P \leqslant 1$) represents the expectation of suppliers about price reversal, in the case when the roots of the auxiliary equation are complex.

7. Discuss the nature of the solution of (76) when the roots of the auxiliary equation are real and distinct, real and coincident or complex conjugate.

8. Discuss the Harrod-Domar growth model

$$Y_t = (1 + v)Y_{t-1} - (v + s)Y_{t-2}$$

where $s = 1 - c =$ marginal propensity to save and v is the power of the accelerator. Discuss also all possible solutions of

$$Y_t = \left(v + \frac{v + s}{v}\right)Y_{t-1} - (v + s)Y_{t-2}$$

5.4 MATHEMATICAL MODELLING THROUGH DIFFERENCE EQUATIONS IN POPULATION DYNAMICS AND GENETICS

5.4.1 Non-Linear Difference Equations Model for Population Growth: Non-Linear Difference Equations

Let x_t be the population at time t and let births and deaths in time-interval $(t, t + 1)$ be proportional to x_t, then the population x_{t+1} at time $t + 1$ is given by

$$x_{t+1} = x_t + bx_t - d x_t = x_t(1 + a) \tag{89}$$

This has the solution

$$x_t = x_0(1 + a)^t, \tag{90}$$

so that the population increases or decreases exponentially according as $a > 0$ or $a < 0$. We now consider the generalisation when births and deaths b and d per unit population depend linearly on x_t so that

$$x_{t+1} = x_t + (b_0 - b_1 x_t)x_t - (d_0 + d_1 x_t)x_t$$

$$= mx_t - rx_t^2 = mx_t \left(1 - \frac{r}{m} x_t\right) \tag{91}$$

This is the simplest non-linear generalisation of (90) and gives the discrete version of the logistic law of population growth. However this model shows many new features not present in the continuous version of the logistic model. Let $rx_t/m = y_t$, then (91) becomes

$$y_{t+1} = my_t(1 - y_t) \tag{92}$$

One-Period Fixed Points and Their Stability

A one-period fixed point of this equation is that value of y_t for which $y_{t+1} = y_t$ i.e. for which

$$y_t = my_t(1 - y_t), \tag{93}$$

so that there are two one-period fixed points 0 and $(m - 1)/m$. If $y_0 = 0$, then $y_1, y_2, y_3, \ldots$ are all zero and the population remains fixed at zero value:

If $y_0 = (m - 1)/m$, then $y_1, y_2, y_3, \ldots$ are all equal to $(m - 1)/m$. The second fixed point exists only if $m > 1$.

We now discuss the stability of equilibrium of each of these equilibrium positions.

Putting $y_t = 0 + u_t$ in (92) and neglecting squares and higher powers of u_t, we get $u_{t+1} = mu_t$ and since $m > 0$, the first equilibrium position is one of unstable equilibrium.

Again putting $y_t = (m - 1)/m + u_t$ in (92) and neglecting squares and higher powers of u_t, we get

$$u_{t+1} = (2 - m)u_t, \tag{94}$$

so that the second position of equilibrium is stable only if $-1 < 2 - m < 1$ or if $1 > m - 2 > -1$ or if $1 < m < 3$.

Thus if $0 < m < 1$, there is only one one-period fixed point and it is unstable. If $1 < m < 3$, there are two one-period fixed points, the first is unstable and the second is stable. If $m > 3$, there are two one-period fixed points, both of which are unstable.

Two-Period Fixed Points and Their Stability

A point is called a two-period fixed point if it repeats itself after two periods i.e. if $y_{t+2} = y_t$ i.e. if

$$y_{t+2} = my_{t+1}(1 - y_{t+1}) = m^2 y_t(1 - y_t)(1 - my_t + my_t^2) = y_t \tag{95}$$

or

$$y_t(my_t - (m-1))(m^2y_t^2 - m(1+m)y_t + (1-m)) = 0 \qquad (96)$$

This is a fourth degree equation and as such there can be four two-period fixed points. Two of these are the same as the one-period fixed points. This is obvious from the consideration that every one-period fixed point is also a two-period fixed point. The genuine two-period fixed points are obtained by solving the equation

$$m^2y_t^2 - m(1+m)y_t + (1+m) = 0 \qquad (97)$$

Its roots are real if $m > 3$. Thus if $m > 3$, the two one-period fixed points become unstable, but two new two-period fixed points exist and we can discuss their stability as before.

It can be shown that if $m_2 < m < m_4$, where $m_2 = 3$ and m_4 is a number slightly greater than 3, then the two two-period fixed points are stable but if $m > m_4$, all the four one- and two-periods become unstable, but four new four-period fixed points exist which are stable if $m_4 < m < m_8$ and become unstable if $m > m_8$.

2^n-Period Fixed Points and Their Stability

It can be shown that there exists an increasing infinite sequence of real numbers $m_2, m_4, m_8, \ldots, m_{2n}, m_{2n+1}, \cdots$ such that when $m_{2n} < m < m_{2n+1}$ there are $2^{n+1}2^{n+1}$-period fixed points, out of which 2^n fixed points are also fixed points of lower order time periods and all these are unstable and the remaining 2^n points are genuine 2^{n+1} period fixed points and are stable.

From 5.3 represents the stable fixed period points.

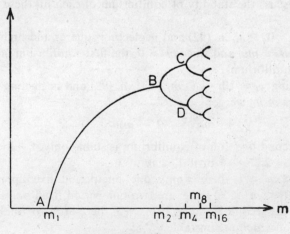

Figure 5.3

When m lies between m_1 and m_2, there is one stable one-period fixed point. When m lies between m_2 and m_4 there are two stable two-period fixed points.

When m lies between m_4 and m_8, there are four stable four-period fixed points, and so on.

Fixed Points of other Periods

The sequence m_2, m_4, m_8, ... is bounded above by a fixed number m^*. If $m > m^*$, there can be a three-period fixed point and if there is a three-period fixed point, there will also be fixed points of periods,

$$3, 5, 7, 9, \ldots$$
$$2 \cdot 3, 2 \cdot 5. \; 2 \cdot 7, 2 \cdot 9, \ldots \tag{98}$$
$$2^2 \cdot 3, 2^2 \cdot 5, 2^2 \cdot 7, \ldots$$

This is expressed by saying that Period Three Means Chaos.

Chaotic Behaviour of the Non-linear Model

If m lies between m_8 and m_{16}, there will be eight 16-period stable fixed points. If a population size starts from any one of these values, it will oscillate through fifteen other values to return to the original value and this pattern will go on repeating itself. If we draw the graph, it will show rapid oscillations and will look like the graph representing a random phenomenon. Our model is perfectly deterministic, though its behaviour may *appear* to be random and stochastic.

Special Features of Non-linear Difference Equation Models

The simple model illustrates the differences in behaviour between difference and differential equation models. The problems of existence and uniqueness of solutions, of the stability of equilibrium positions are all different due to the basic fact that inspite of similarities, the Discrete and the Continuous are really different.

5.4.2 Age-Structured Population Models

Let $x_1(t)$, $x_2(t)$, ..., $x_p(t)$ be the population sizes of p pre-reproductive age-groups at time t;

Let $x_{p+1}(t)$, $x_{p+2}(t)$, ..., $x_{p+q}(t)$ be the population sizes of q reproductive age-groups at time t, and

Let $x_{p+q+1}(t)$, $x_{p+q+2}(t)$, ..., $x_{p+q+r}(t)$ be the population sizes of r post-reproductive age-groups at time t.

Let b_{p+1}, b_{p+2}, ..., b_{p+q} be the birth rates i.e. the number of births per unit time per individual in the reproductive age groups.

In other age-groups, the birth rates are zero.

Let d_1, d_2, ..., d_{p+q+r} be the death rates in the $p + q + r$ age-groups.

Let m_1, m_2, ..., m_{p+q+r}, be the rates of migration to the next age-groups, then we get the system of difference equations

$$x_1(t + 1) = b_{p+1}x_{p+1}(t) + \ldots + b_{p+q}x_{p+q}(t) - (d_1 + m_1)x_1(t)$$
$$x_2(t + 1) = m_1x_1(t) - (d_2 + m_2)x_2(t)$$
$$\ldots \qquad \ldots \qquad \ldots \tag{99}$$
$$x_{p+q+r-1}(t + 1) = m_{p+q+r-2}(t) - (d_{p+q+r-1} + m_{p+q+r-1})x_{p+q+r-1}(t)$$
$$x_{p+q+r}(t + 1) = m_{p+q+r-1}x_{p+q+r-1}(t) - (d_{p+q+r})x_{p+q+r}(t)$$

which can be written in the matrix form

$$X(t + 1) = LX(t),\qquad\qquad(100)$$

where

$$X(t) = \begin{bmatrix} x_1(t) \\ x_2(t) \\ \cdot \\ \cdot \\ \cdot \\ x_{p+q+r}(t) \end{bmatrix},$$

$$L = \begin{bmatrix} -(d_1 + m_1) & 0 & 0 \ldots 0 & b_{p+1} & b_{p+2} \ldots b_{p+q} & 0 & .. & 0 & 0 \\ m_1 & -(d_2 + m_2) & 0 \ldots & 0 & 0 & \ldots & 0 & 0 & .. & 0 & 0 \\ 0 & m_2 & -(d_3 + m_3) & 0 & 0 & \ldots & 0 & 0 & .. & 0 & 0 \\ \cdots & \cdots & \cdots & \cdots & \cdots & & \cdot & & & \cdot & \cdot \\ \cdots & \cdots & \cdots & \cdots & \cdots & & \cdot & & & \cdot & \cdot \\ \cdots & \cdots & \cdots & \cdots & \cdots & & \cdot & & & \cdot & \cdot \\ 0 & 0 & 0 \ldots & 0 & 0 & & 0 & 0 & .. & m_{n-1} & -d_n \end{bmatrix}$$

$$(101)$$

where $p + q + r = n$.

L is called the Leslie matrix. All the elements of its main diagonal are negative and all the elements of its main subdiagonal are positive. In addition q elements in the first row are positive and the rest of the elements are all zero. The solution of (100) can be written as

$$X(t) = L^t X(0)\qquad\qquad(102)$$

Now the Leslie matrix has the property that it has a dominant eigenvalue which is real and positive, which is greater in absolute value than any other eigenvalue and for which the corresponding eigenvector has all its components positive. If this dominant eigenvalue is greater than unity, then the populations of all age-groups will increase exponentially and if it is less than unity the population of all age-groups will die out. If this dominant eigenvalue is unity, the population can have a stable age structure.

The Leslie model is in terms of a system of linear difference equations. If we take the effects of overcrowding and density dependence into account, the equations are nonlinear.

5.4.3 Mathematical Modelling through Difference Equations in Genetics

(a) Hardy-Weinberg Law

Every characteristic of an individual, like height or colour of the hair, is determined by a pair of genes, one obtained from the father and the other obtained from the mother. Every gene occurs in two forms, a dominant (denoted by a capital letter say G) and a recessive (denoted by the corresponding small letter say g). Thus with respect to a characteristic, an individual may be a dominant (GG), a hybrid (Gg or gG) or a recessive (gg).

In the nth generation, let the proportions of dominants, hybrids and recessives be p_n, q_n, r_n so that

$$p_n + q_n + r_n = 1, \qquad p_n \geqslant 0, q_n \geqslant 0, r_n \geqslant 0 \tag{103}$$

We assume that individuals, in this generation mate at random. Now p_{n+1} = the probability that an individual in the $(n + 1)$th generation is a dominant (GG) = (probability that this individual gets a G from the father) $\times$ (probability that the individual gets a G from the mother)

$$= \left(p_n + \frac{1}{2}q_n\right)\left(p_n + \frac{1}{2}q_n\right) = \left(p_n + \frac{1}{2}q_n\right)^2$$

or

$$p_{n+1} = \left(p_n + \frac{1}{2}q_n\right)^2 \tag{104}$$

Similarly

$$q_{n+1} = 2\left(p_n + \frac{1}{2}q_n\right)\left(r_n + \frac{1}{2}q_n\right) \tag{105}$$

$$r_{n+1} = \left(r_n + \frac{1}{2}q_n\right)^2, \tag{106}$$

so that

$$p_{n+1} + q_{n+1} + r_{n+1} = \left(p_n + \frac{1}{2}q_n + \frac{1}{2}q_n + r_n\right)^2 = 1, \tag{107}$$

as expected. Similarly

$$p_{n+2} = \left(p_{n+1} + \frac{1}{2}q_{n+1}\right)^2$$

$$= \left(\left(p_n + \frac{1}{2}q_n\right)^2 + \left(p_n + \frac{1}{2}q_n\right)\left(r_n + \frac{1}{2}q_n\right)\right)^2$$

$$= \left(p_n + \frac{1}{2}q_n\right)^2 \left(p_n + \frac{1}{2}q_n + \frac{1}{2}q_n + r_n\right)^2$$

$$= \left(p_n + \frac{1}{2}q_n\right)^2 = p_{n+1} \tag{108}$$

and

$$q_{n+2} = q_{n+1}, \qquad r_{n+2} = r_{n+1}, \tag{109}$$

so that the proportions of dominants, hybrids and recessives in the $(n + 2)$th generation are same as in the $(n + 1)$th generation.

Thus in any population in which random mating takes place with respect to a characteristic, the proportions of dominants, hybrids and recessive do not change after the first generation. This is known as Hardy-Weinberg law after the mathematician Hardy and geneticist Weinberg who jointly discovered it.

The equations (104)–(107) is a set of difference equations of the first order.

(b) *Improvement of Plants through Elimination of Recessives*

Suppose the recessives are undesirable and as such we do not allow the recessives in any generation to breed.

Let p_n, q_n, r_n be the proportions of dominants, hybrids and recessives before elimination of recessives and let p'_n, q'_n, 0 be the populations after the elimination, then

$$\frac{p'_n}{p_n} = \frac{q'_n}{q_n} = \frac{p'_n + q'_n}{p_n + q_n} = \frac{1}{1 - r_n} \tag{110}$$

Now we allow random mating and let p_{n+1}, q_{n+1}, r_{n+1} be the proportions in the next generation before elimination of recessives, then using (104)–(108)

$$p_{n+1} = \left(p'_n + \frac{1}{2} q'_n \right)^2 \tag{111}$$

$$q_{n+1} = 2\left(p'_n + \frac{1}{2} q'_n \right)\left(\frac{1}{2} q'_n \right) = q'_n \left(p'_n + \frac{1}{2} q'_n \right) \tag{112}$$

$$r_{n+1} = \left(\frac{1}{2} q'_n \right)^2 = \frac{1}{4} q'^2_n \tag{113}$$

After elimination of recessives, let the new proportions be p'_{n+1}, q'_{n+1}, so that

$$\frac{p'_{n+1}}{p_{n+1}} = \frac{q'_{n+1}}{q_{n+1}} = \frac{1}{p_{n+1} + q_{n+1}} = \frac{1}{1 - \frac{1}{4} q'^2_n} \tag{114}$$

so that

$$q'_{n+1} = \frac{q'_n(p'_n + \frac{1}{2} q'_n)}{1 - \frac{1}{4} q'^2_n} = \frac{q'_n(1 - \frac{1}{2} q'_n)}{1 - \frac{1}{4} q'^2_n}$$

$$= \frac{q'_n}{1 + \frac{1}{2} q'_n} \tag{115}$$

This is a non-linear difference equation of the first order. To solve it we substitute

$$q'_n = 1/u_n$$

to get

$$u_{n+1} = u_n + \frac{1}{2} \tag{116}$$

which has the solution

$$u_n = A + \frac{1}{2} n \tag{117}$$

or

$$q'_n = \frac{1}{A + \frac{1}{2} n} \tag{118}$$

so that $q'_n \to 0$ and $p'_n \to 1$ as $n \to \infty$. Thus ultimately we should be left with all dominants. Equation (118) determines the rate at which hybrids disappear.

EXERCISE 5.4

1. Show that in Figure 5.3, AB is the arc of a rectangular hyperbola.
2. Find m_4 and draw the curves BC and BD.

3. Find the four stable eight-period fixed points.

4. For the condition for the existence of a three-period fixed point.

5. Find the characteristic equation for the Leslie matrix and show that it always has a positive real root. Find the condition that this root is less than unity.

6. Let $y_{t+1} = 3.1(1 - y_t)$. Draw the graph of its solution for $y_0 = 0.5$.

7. Draw the graphs of $\ln x_1(t)$, $\ln x_2(t)$, $\ln x_3(t)$ for the system

$$X(t + 1) = AX(t) \quad \text{when}$$

$$A = \begin{bmatrix} 0 & 10 & 8 \\ \frac{1}{3} & 0 & 0 \\ 0 & \frac{1}{2} & 0 \end{bmatrix} \text{ or } \begin{bmatrix} 0 & 2 & 2 \\ \frac{1}{3} & 0 & 0 \\ 0 & \frac{1}{2} & 0 \end{bmatrix} \text{ or } \begin{bmatrix} 0 & \frac{1}{2} & \frac{1}{4} \\ \frac{1}{3} & 0 & 0 \\ 0 & \frac{1}{2} & 0 \end{bmatrix}$$

when $x_1(0) = 10$, $x_2(0) = 10$, $x_3(0) = 10$
and interpret the graphs.

8. Discuss the problem of Section 5.4.3(b) when only a fraction k of the recessives are eliminated at each stage.

5.5 MATHEMATICAL MODELLING THROUGH DIFFERENCE EQUATIONS IN PROBABILITY THEORY

5.5.1 Markov Chains

Let a system be capable of being in n possible states $1, 2, \ldots, n$ and let the probability of transition from state i to state j in time interval t to $t + 1$ be p_{ij}. Let $p_j(t)$ denote the probability that the system is in state j at time t $(j = 1, 2, \ldots, n)$, then at time $t + 1$ it can be in any one of the states $1, 2, \ldots, n$.

It can be in the ith state at time $t + 1$ in n exclusive ways since it could have been in any one of the n states $1, 2, \ldots, n$ at time t and it could have transited from that state to ith state in time interval $(t, t + 1)$. By using the theorems of total and compound probability, we get

$$p_i(t + 1) = \sum_{j=1}^{n} p_{ji}p_j(t), \qquad i = 1, 2, \ldots, n \tag{119}$$

or
$$
\begin{aligned}
p_1(t + 1) &= p_{11}p_1(t) + p_{21}p_2(t) + \ldots + p_{n1}p_n(t) \\
p_2(t + 1) &= p_{12}p_1(t) + p_{22}p_2't) + \ldots + p_{n2}p_n(t) \\
&\cdots \quad\quad \cdots \quad\quad \cdots \quad\quad \cdots \quad\quad \cdots \\
p_n(t + 1) &= p_{1n}p_1(t) + p_{2n}p_2(t) + \ldots + p_{nn}p_n(t)
\end{aligned}
\tag{120}
$$

or
$$
\begin{bmatrix} p_1(t + 1) \\ p_2(t + 1) \\ \cdot \\ \cdot \\ p_n(t + 1) \end{bmatrix} = \begin{bmatrix} p_{11} & p_{21} & \cdots & p_{n1} \\ p_{12} & p_{22} & \cdots & p_{n2} \\ \cdots & \cdots & \cdots & \cdots \\ p_{1n} & p_{2n} & \cdots & p_{nn} \end{bmatrix} \begin{bmatrix} p_1(t) \\ p_2(t) \\ \cdot \\ \cdot \\ p_n(t) \end{bmatrix}
\tag{121}
$$

or
$$P(t + 1) = AP(t),\qquad(122)$$

where $P(t)$ is a probability vector and A is a matrix, all of whose elements lie between zero and unity (since these are all probabilities). Further the sum of elements of every column is unity, since the sum of elements of the ith column is $\overset{n}{\underset{j=1}{\Sigma}}\ p_{ij}$ as this denotes the sum of the probabilities of the system going from the ith state to any other state and this sum must be unity.

The solution of the matrix difference equation (122) is

$$P(t) = A^t P(0)\qquad(123)$$

If all the eigenvalues $\lambda_1, \lambda_2, \ldots, \lambda_n$ of A are distinct, we can write

$$A = S\Lambda S^{-1}\qquad(124)$$

where
$$\Lambda = \begin{bmatrix} \lambda_1 & 0 & 0 & \ldots & 0 \\ 0 & \lambda_2 & 0 & \ldots & 0 \\ \ldots & \ldots & \ldots & \ldots & \ldots \\ 0 & 0 & 0 & \ldots & \lambda_n \end{bmatrix}\qquad(125)$$

so that
$$A^t = (S\Lambda S^{-1})(S\Lambda S^{-1}) \ldots (S\Lambda S^{-1})$$
$$= S\Lambda^t S^{-1}$$

$$= S \begin{bmatrix} \lambda_1^t & 0 & 0 & \ldots & 0 \\ 0 & \lambda_2^t & 0 & \ldots & 0 \\ \ldots & \ldots & \ldots & \ldots & \ldots \\ 0 & 0 & 0 & \ldots & \lambda_n^t \end{bmatrix} S^{-1}\qquad(126)$$

The probability vector will not change if $P(t + 1) = P(t)$ so that from (122)

$$(I - A)P(t) = 0\qquad(127)$$

Thus if P is the eigenvector of the matrix A corresponding to unit eigenvalue, then P does not change i.e. if the system start with probability vector P at time 0, it will always remain in this state. Even if the system starts from any other probability vector, it will ultimately be described by the probability vector P as $t \to \infty$.

As a special case, suppose we have a machine which can be in two states, working or non-working. Let the probability of its transition from working to non-working be α, of its transition from non-working to working be β, then the transition probability matrix A is obtained from

$$\begin{array}{cc} & \text{working} \quad \text{non-working} \\ \begin{array}{c} \text{working} \\ \text{non-working} \end{array} & \begin{bmatrix} 1 - \alpha & \alpha \\ \beta & 1 - \beta \end{bmatrix} \end{array}\qquad(128)$$

The system of difference equations is

$$p_1(t + 1) = p_1(t)(1 - \alpha) + p_2(t)\beta$$
$$p_2(t + 1) = p_1(t)\alpha + p_2(t)(1 - \beta) \tag{129}$$

or

$$\begin{bmatrix} p_1(t + 1) \\ p_2(t + 1) \end{bmatrix} = \begin{bmatrix} 1 - \alpha & \beta \\ \alpha & 1 - \beta \end{bmatrix} \begin{bmatrix} p_1(t) \\ p_2(t) \end{bmatrix} \tag{130}$$

The eigenvalues of the matrix A is given by

$$\begin{vmatrix} 1 - \alpha - \lambda & \beta \\ \alpha & 1 - \beta - \lambda \end{vmatrix} = 0 \text{ or } (\lambda - 1)(\lambda - \overline{1 - \alpha - \beta}) = 0 \tag{131}$$

The eigenvector corresponding to the unit eigenvalue is $\beta/(\alpha + \beta)$, $\alpha/(\alpha + \beta)$ and as such ultimately the probability of the machines being found in working order is $\beta/(\alpha + \beta)$ and the probability of its being found in a non-working state is $\alpha/(\alpha + \beta)$.

5.5.2 Gambler's Ruin Problems

Let a gambler with capital n dollars play against an infinitely rich adversary. Let the probability of his winning or losing a unit dollar in any game be p and q respectively where $p + q = 1$ and let p_n be the probability of his being ultimately ruined. At the next game, the probability of his winning is p and if he wins, his capital would become $n + 1$ and the probability of his ultimate ruin would be p_{n+1}. On the other hand if he loses at the next game, the probability for which is q, his capital would become $n - 1$ and the probability of his ultimate ruin would be p_{n-1}, so that we get the linear difference equation of the second order

$$p_n = pp_{n+1} + qp_{n-1} \tag{132}$$

The auxiliary equation for this is

$$p\lambda^2 - \lambda + (1 - p) = 0$$

or

$$p(\lambda - 1)\left(\lambda - \frac{1 - p}{p}\right) = 0 \tag{133}$$

As such the solution of (132) is

$$p_n = A + B\left(\frac{q}{p}\right)^n \tag{134}$$

Now let the gambler decide to stop this game when his capital becomes a dollars so that the probability of his being ruined when his starting capital is a dollars is zero i.e. $p_a = 0$. In the same way when his starting capital is zero, he is already ruined, so we put $p_0 = 1$. Using

$$p_0 = 1, \qquad p_a = 0 \tag{135}$$

(134) gives

$$p_n = \frac{(q/p)^a - (q/p)^n}{(q/p)^a - 1} \tag{136}$$

Now let D_n denote the expected number of games before the gambler is ruined. If he wins at the next game, his capital becomes $n + 1$ and the expected number of games would then be D_{n+1} and if he loses, his capital becomes $n - 1$ and the expected number of games would be only D_{n-1}. As such, we get

$$D_n = pD_{n+1} + qD_{n-1} + 1 \tag{137}$$

with boundary conditions

$$D_0 = 0, \quad D_a = 0 \tag{138}$$

This gives the solution

$$D_n = \frac{n}{q - p} - \frac{a}{q - p} \frac{1 - (q/p)^n}{1 - (q/p)^a}$$

EXERCISE 5.5

1. Show that the solution of (129) is

$$p_1(t) = \frac{\beta}{\alpha + \beta} + (1 - \alpha - \beta)^t \left(p_1(0) - \frac{\beta}{\alpha + \beta} \right)$$

$$p_2(t) = \frac{\alpha}{\alpha + \beta} + (1 - \alpha - \beta)^t \left(p_2(0) - \frac{\alpha}{\alpha + \beta} \right)$$

2. Show that $-1 < 1 - \alpha - \beta < 1$ and deduce that $p_1(t) \to \dfrac{\beta}{\alpha + \beta}$ and $p_2(t) \to \dfrac{\alpha}{\alpha + \beta}$ as $t \to \infty$. Show also that $\beta/(\alpha + \beta)$, $\alpha/(\alpha + \beta)$ give the components of the eigenvector of the matrix A corresponding to the unit eigenvalue.

3. In a panel survey, a person gives an answer 'yes' or 'no'. The probability of his changing from 'yes' to 'no' in the next survey is α and that of changing from 'no' to 'yes' is β. Find the probability that ultimately he will answer 'yes'.

4. In a game of chance, the probability of a person winning a second game after losing the first game is α and the probability of his losing a second game after winning the first game is β. Find the ultimate chance of winning.

5. Show that if $p = q = 1/2$, the solution of (132) is

$$p_n = 1 - n/a$$

Show also that this is the limiting value of p_n given by (136) when p and q both approach $1/2$.

6. Show that if $p = q = \frac{1}{2}$, the solution of (137) is

$$D_n = n(a - n)$$

Show also that this is the limiting value of D_n given by (139) when p and q both approach $1/2$.

7. In gambler's ruin problem, discuss the special cases when

$$n = 1 \quad \text{or} \quad n = a - 1.$$

8. A particle is at the point n on the positive real axis where n is a non-negative integer. At every unit interval of time it can move unit distance towards the right or towards the left with probability p and $q(p + q = 1)$ respectively. If the particle reaches 0 or a, it is absorbed there. Find the probabilities of the particle being ultimately absorbed at 0 or at a. Find also the expected duration before absorption in either case.

9. n letters to each of which corresponds an envelope are placed in the envelopes at random. If u_n is the number of ways in which all letters go wrong, show that

$$u_n = (n - 1)(u_{n-1} + u_{n-2})$$

Prove that $u_n - nu_{n-1} = (-1)^{n-2}(u_2 - 2u_1) = (-1)^n$

and $$u_n = n! \left[\frac{1}{2!} - \frac{1}{3!} + \ldots + \frac{(-1)^n}{n!}\right]$$

Deduce that the probability that all n letters go wrong is given by the first $(n - 1)$ terms in expression of $1 - e^{-1}$.

10. A player tosses a coin and is to score one point for every head turned up and two for every tall. He is to play on until his score reaches or passes n. If p_n is the probability of attaining exactly n, show that

$$p_n = \frac{1}{2}(p_{n-1} + p_{n-2}), \quad p_n = \frac{1}{2}\left[2 + (-1)^n \frac{1}{2^n}\right].$$

5.6 MISCELLANEOUS EXAMPLES OF MATHEMATICAL MODELLING THROUGH DIFFERENCE EQUATIONS

Difference equations arise in economics since values of prices, quantities, national income, savings, investments at discrete intervals of time are related. These arise in genetics because proportions of dominants, hybrids and recessives in different generations are related by genetic laws. These arise in population dynamics because population sizes at discrete instants of time are related by births, deaths, immigration and emigration. These arise in finance because amounts at discrete instants of time are related by rates of interest. These arise in gambler's ruin problem because the probability of ruin (or duration of game) when gambler's capital is n is related to the probability of ruin (or duration of game) when his capital is $n + 1$.

Similarly in geometry, difference equations can arise because the number of compartments in which n lines or curves divide a plane or surface is related to the number of components determined by $(n + 1)$ lines or curves; in dynamics the ranges after successive rebounds of an elastic ball from a horizontal or inclined place are related; in electrical currents, the potential at neighbouring nodes and currents in neighbouring circuits are related by Kirchhoff's laws and so on.

EXERCISE 5.6

1. If u_n is the number of compartments formed by n straight lines drawn in a plane such that no two are parallel and no three are concurrent, show that

$$u_{n+1} = u_n + (n + 1), \qquad u_n = \frac{1}{2}(n^2 + n + 2).$$

2. Show that if u_n is the number of compartments formed when n closed curves are drawn on a closed surface in such a way that no three intersect at the same point and every pair crosses at two points and only at two points then

$$u_n = u_n + 2n, \qquad u_n = n^2 - n + 2$$

3. If $I_n = \displaystyle\int_0^\pi \frac{\cos n\theta \, d\theta}{\cos \theta - \cos \alpha}$, show that $I_n + I_{n-2} = 2 \cos \alpha \, I_{n+1}$ and hence show that $I_n = \pi \sin n\alpha / \sin \alpha$.

4. Using the difference equation

$$(n + 1)P_{n+1}(x) - (2n + 1)xP_n(x) + nP_{n-1}(x) = 0$$

valid for Legender's polynomials, evaluate

$$I_n = \int_{-1}^1 \frac{P_n(x)P_{n-1}(x)}{x} \, dx$$

by first showing that

$$(n + 1)I_{n+1} + nI_n = 2.$$

5. N equal uniform rods, smoothly jointed together and at rest in a straight line on a horizontal table, have an impulse J applied to the free end of the first rod, J being horizontal and perpendicular to the line or rods. Denoting the equal and opposite reactions at the ith joint by R_i, and adopting the convention that the impulse R_i acting on the $(i + 1)$th rod is measured in the same sense as J, prove that

$$R_{i-1} + 4R_i + R_{i+1} = 0$$

and explain what values have to be given to R_0 and R_N in order to make the equation hold for $i = 1, 2, \ldots, N - 1$.

6. Fibonacci's numbers are defined by $F_1 = 1, F_2 = 1, F_n = F_{n-1} + F_{n-2}$; find F_n and an asymptotic formula for it when n is large.

7. Generalised Fibonacci's numbers are defined by

$$F_{n,r} = F_{n-1,r} + F_{n-2,r} + \ldots + F_{n-r,r}$$

Find formula for $F_{n,r}$ and discuss its properties.

8. In the steady-state, the probability of there being n persons in a queue is given by

$$(\lambda + \mu)p_n = \lambda p_{n-1} + \mu p_{n+1}, \qquad n = 0, 1, 2, 3, \ldots$$

show that $\qquad p_n = (1 - \rho)\rho^n; \qquad \rho = \lambda/\mu.$

9. Show that the number of transformation of n points into themselves in which $n - r$ points remain fixed is given by

$$^nc_r \, r! \left(\frac{1}{2!} - \frac{1}{3!} + \frac{1}{4!} \cdots + \frac{(-1)^r}{r!} \right)$$

10. Show that the number of transformations in which no point remains fixed and in which just one point remain fixed differ always by unity

Mathematical Modelling Through Partial Differential Equations

6.1 SITUATIONS GIVING RISE TO PARTIAL DIFFERENTIAL EQUATION MODELS

Partial differential equation (PDE) models arise when the variables of interest are functions of more than one independent variable and all the dependent and independent variables are continuous. Thus in fluid dynamics, the velocity components u, v, w and the pressure p at any point x, y, z and at any time t are functions of x, y, z, t and in general $u(x, y, z, t)$, $v(x, y, z, t)$, $w(x, y, z, t)$, $p(x, y, z, t)$ are continuous functions, with continuous first and second order partial derivatives, of the continuous independent variables x, y, z, t. Similarly the electric field intensity vector $\vec{E}(x, y, z, t)$, the magnetic field intensity vector $\vec{H}(x, y, z, t)$, the electric current density vector $\vec{J}(x, y, z, t)$, the temperature $T(x, y, z, t)$ and the displacement vector $\vec{D}(x, y, z, t)$, of an elastic substance are in general continuous vector or scalar functions with continuous derivatives. One object of mathematical modelling is to translate the physical laws governing these functions into partial differential equations whose solution, subject to appropriate initial and boundary conditions, should determine the values of these functions at any point x, y, z at any time t. For this purpose, we consider an elemetary volume element and apply to it the principles of continuity and heat, momentum, energy balance etc.

According to the principle of mass balance, the amount of the substance flowing across the surface of the volume element in a small time Δt is equal to the decrease in the mass of the substance inside the volume in that time. The amount of the mass flowing across the surface can be expressed as a surface integral and the change of mass inside the volume can be expressed as a volume integral. However the surtace integral can also be converted into a volume integral by using Gauss divergence theorem so that finally the mass balance principle requires the vanishing of a volume integral for all arbitrary volume elements. This can happen only if the integrand vanishes identically. The vanishing of the integrand gives rise to a partial differential equation. We shall discuss this method of deriving partial differential equations in Section 6.2.

Here we have applied the principle of mass balance on a global basis i.e. to any volume element, large or small. However the procedure finally gives a partial differential equation valid locally at every point of the region concerned.

If we apply the momentum-balance principle in the form of Newton's second law viz. that the mass of a volume element multiplied by its acceleration vector is equal to the vector sum of all the external body forces acting on the volume element and the internal forces due to the action of the rest of the substance on the volume element under consideration, we get directly a partial differential equation. We shall discuss the derivation of these partial differential equations in Section 6.3.

Partial differential equations also arise due to application of variational principles of science and engineering. These require us to choose $u(x, y, z, t)$, $v(x, y, z, t)$, $w(x, y, z, t)$ etc. as functions of x, y, z, t so as to maximize or minimize the integral of a known function $F(x, y, z, t, u, v, w, u_x, u_y, u_z, u_t, \ldots)$. This is achieved by solving Euler-Lagrange equations of calculus of variations. These equations are partial differential equations. This third method of mathematical modelling through P.D.E. will be discussed in Section 6.6.

Sometimes partial differential equations can also be useful when the independent variables are not all continuous. Thus let, $p(m, n, t)$ be the probability of there being m susceptibles and n infected persons at time t in an epidemic area, then we cannot get a partial differential equation for $p(m, n, t)$ since m and n are discrete integer-values variables. However if we define the probability generating function

$$\Phi(u, v, t) = \sum_{n=0}^{\infty} \sum_{m=0}^{\infty} p(m, n, t) \, u^m v^n, \tag{1}$$

then we can possibly get a p.d.e for $\Phi(u, v, t)$ since u, v are continuous Solving we can get $\Phi(u, v, t)$ and expanding this function in powers of u, v, we can get $p(m, n, t)$ for all values of m, n and t.

EXERCISE 6.1

1. Use Divergence Theorem to evaluate

(a) $\iint\limits_{S} x \, dy \, dz + y \, dz \, dx + z \, dx \, dy$; $S: x^2 + y^2 + z^2 = a^2$

(b) $\iint\limits_{S} x^2 \, dy \, dz + y^2 \, dz \, dx + z^2 \, dx \, dy$; $S:$ Surface bounding $0 \leqslant x, y, z \leqslant a$

2. Use Divergence Theorem to show that

(a) $\iint\limits_{S} \text{curl } \vec{F} \cdot \vec{dS} = 0$

(b) $\iint\limits_S (f\,\nabla g - g\nabla f)\cdot d\vec{S} = \iint\limits_S \left(f\,\dfrac{\partial g}{\partial n} - g\,\dfrac{\partial f}{\partial n}\right) dS = 0$

3. Use Divergence Theorem to show that the volume V of a region T bounded by a surface s is given by

$$V = \iint\limits_S x\,dy\,dz = \iint\limits_S y\,dz\,dx = \iint\limits_S z\,dx\,dy$$

$$= \frac{1}{3}\iint\limits_S (x\,dy\,dz \;\; +y\,dz\,dx + z\,dx\,dy)$$

Verify these formula for a sphere.

4. Find the probability generating functions for the following distributions.

(a) Binomial Distribution $\quad P(r) = {}^n c_r\,p^r\,q^{n-r}, \quad r = 0, 1, 2, \ldots, n$

(b) Poisson Distribution: $\quad P(r) = e^{-m}\,m^r/r!; \quad r = 0, 1, 2, 3, \ldots$

(c) Geometric Distribution $\quad P(r) = q^r p \quad\;\; ; \quad r = 0, 1, 2, 3, \ldots$

6.2 MASS-BALANCE EQUATIONS: FIRST METHOD OF GETTING PDE MODELS

6.2.1 Equation of Continuity in Fluid Dynamics

If V_n is the normal component of the velocity of the fluid at any point of the surface of our conceptual volume element (Fig. 6.1), the mass of the fluid flowing out in time Δt across the surface

$$= \Delta t \iint\limits_S \rho V_n\,dS = \Delta t \iint\limits_S \rho\vec{V}\cdot d\vec{S}$$

$$= \Delta t \iiint\limits_T \text{div}\,(\rho\,\vec{V})\,dx\,dy\,dz, \quad (2)$$

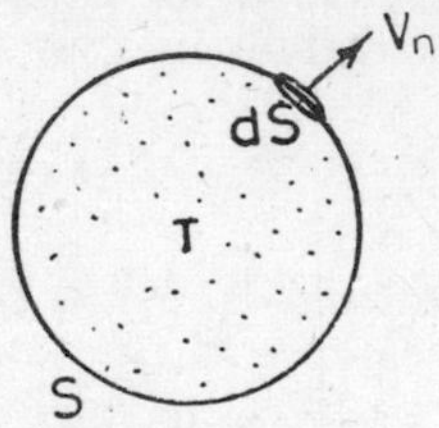

on using Gauss's Divergence Theorem. The change of mass of fluid in the volume element in the time Δt is given by

Figure 6.1

$$-\Delta t\,\frac{\partial}{\partial t}\iiint\limits_T \rho\,dx\,dy\,dz = -\Delta t\iiint\limits_T \frac{\partial\rho}{\partial t}\,dx\,dy\,dz \tag{3}$$

Using (2) and (3), the principle of mass-balance gives

$$\iiint\limits_T\left[\frac{\partial\rho}{\partial t} - \text{div}\,(\rho\,\vec{V})\right] dx\,dy\,dz = 0 \tag{4}$$

Since (4) is to be true for all arbitrary volume elements, we get

$$\frac{\partial\rho}{\partial t} + \text{div}\,(\rho\,\vec{V}) = 0 \tag{5}$$

or
$$\frac{\partial \rho}{\partial t} + \frac{\partial}{\partial x}(\rho u) + \frac{\partial}{\partial y}(\rho v) + \frac{\partial}{\partial z}(\rho w) = 0 \qquad (6)$$

If the fluid is incompressible, ρ is constant and (5), (6) give

$$\text{div}\,(\vec{V}) = 0 \qquad \text{or} \qquad \frac{\partial u}{\partial x} + \frac{\partial v}{\partial y} + \frac{\partial w}{\partial z} = 0 \qquad (7)$$

Further if the flow is irrotational i.e. if there exists a scalar velocity potential function Φ such that

$$\vec{V} = -\text{grad}\,\Phi \qquad \text{or} \qquad u = -\frac{\partial \Phi}{\partial x},$$

$$v = -\frac{\partial \Phi}{\partial y}, \qquad\qquad w = -\frac{\partial \Phi}{\partial z}, \qquad (8)$$

then (7) and (8) give

$$\Delta^2 \Phi = 0 \qquad \text{or} \qquad \frac{\partial^2 \Phi}{\partial x^2} + \frac{\partial^2 \Phi}{\partial y^2} + \frac{\partial^2 \Phi}{\partial z^2} = 0 \qquad (9)$$

Thus the velocity potential for irrotational flow statisfies the Laplace equation and is a harmonic function.

6.2.2 Equation of Continuity for Heat Flow

In this case, the amount of heat flow across the surface of a volume per unit time is equal to the rate of decrease of heat inside the volume so that (Fig. 6.1)

$$\iint_S V_n \, dS = \iint_S \vec{V} \cdot d\vec{S} = -\frac{\partial}{\partial t} \iiint_T \sigma \rho T \, dx \, dy \, dz, \qquad (10)$$

where $\vec{V}$ is the heat flow velocity, ρ is the density, σ is the specific conductivity and T is the temperature of the substance. Now from physical experiments

$$\vec{V} = -k\boldsymbol{\nabla} T, \qquad (11)$$

where k is the diffusivity of the substance. Assuming σ and ρ to be constant, we get

$$\sigma \rho \iiint_T \frac{\partial T}{\partial t} \, dx \, dy \, dz = \iint_S k\boldsymbol{\nabla} T \cdot d\vec{S} = \iiint_T \text{div}\,(k\boldsymbol{\nabla} T) \, dx \, dy \, dz \quad (12)$$

Since this is true for all volume elements,

$$\sigma \rho \frac{\partial T}{\partial t} = \text{div}\,(k\boldsymbol{\nabla} T)$$

$$= \frac{\partial}{\partial x}\left(k\,\frac{\partial T}{\partial x}\right) + \frac{\partial}{\partial y}\left(k\,\frac{\partial T}{\partial z}\right) + \frac{\partial}{\partial z}\left(k\,\frac{\partial T}{\partial z}\right) \qquad (13)$$

If k is also constant, we get

$$\boldsymbol{\nabla}^2 T = \frac{\sigma \rho}{k}\,\frac{\partial T}{\partial t} \qquad \text{or} \qquad \frac{\partial^2 T}{\partial x^2} + \frac{\partial^2 T}{\partial y^2} + \frac{\partial^2 T}{\partial z^2} = \frac{\sigma \rho}{k}\,\frac{\partial T}{\partial t}$$

This is called the *heat-conduction equation* or the *diffusion equation*. In the steady case i.e. when there is no variation with time, it reduces to Laplace's equation (9).

6.2.3 Equation of Continuity for Traffic Flow on a High-way

Let $P(x, t)$ and $u(x, t)$ be respectively the traffic density (number of cars per unit length of the high way) and velocity of a car on a high way at a distance x from the origin at time t, then if no cars enter or leave the highway, using the continuum model, we get the continuity equation

$$\frac{\partial \rho}{\partial t} + \frac{\partial}{\partial x}(\rho u) = 0 \tag{15}$$

There are two dependent variables viz. $P(x, t)$ and $u(x, t)$ and there is only one equation connecting them. If we can get one more relation between $P(x, t)$ and $u(x, t)$ either empirically or theoretically, we can solve for both $\rho(x, t)$ and $u(x, t)$. We shall discuss this model further in Section 6.5.

6.2.4 Gauss Divergence Theorem in Electrostatics

According to this theorem, the surface integral of $\vec{E}(x, y, z, t)$ over a closed surface is equal to 4π times the electric charge inside the volume enclosed by the surface so that

$$\iint_S \vec{E} \cdot d\vec{S} = 4\pi \iiint_T \rho \, dx \, dy \, dz \tag{16}$$

or
$$\text{div } \vec{E} = 4\pi\rho \tag{17}$$

Since in electrostatics

$$\text{Curl } \vec{E} = 0, \tag{18}$$

there exists an electrostatic potential function Φ such that

$$\vec{E} = -\text{grad } \Phi \tag{19}$$

From (17) and (19)

$$\text{div (grad } \Phi) = -4\pi\rho \quad \text{ or } \quad \nabla^2\Phi = -4\pi\rho, \tag{20}$$

which is called *Poisson's equation*. If $\rho = 0$, i.e. if there is no charge at a point, this reduces to Laplace's equation (9).

6.2.5 Mathematical Modelling in Terms of Laplace's Equation

Laplace equation $\nabla^2\Phi = 0$ provides an appropriate mathematical model for various quantities of interest in physics:

(i) The gravitational potential Φ satisfies Laplace equation (9) in empty space and Poisson equation (20) at a point where there is gravitational matter of density ρ. The force of attraction $\vec{F}$ is then given by

$$\vec{F} = \text{grad } \Phi \tag{21}$$

(ii) At all points of a perfect fluid where there are no sources and sinks and the motion is irrotational, the velocity potential satisfies Laplace's equation (9) and the velocity vector is given by

$$\vec{V} = --\text{grad } \Phi \tag{22}$$

(iii) The electrostatic potential Φ satisfies Laplace's equation (9) at all points in empty space and satisfies Poisson's equation (20) at a point where the density of electric charge is ρ.

(iv) In the presence of dielectrics, the electrostatic potential satisfies the modified Poisson's equation.

$$\text{div } (k \text{ grad } \Phi) = -4\pi\rho, \tag{23}$$

where k is the dielectric permeability. If $\rho = 0$ and k is constant, (23) reduces to Laplace's equation,

(v) The magnetostatic potential Φ satisfies the equation

$$\text{div } (\mu \text{ grad } \Phi) = 0, \tag{24}$$

where σ is the magnetic permeability and the magnetic vector $\vec{H}$ is given by

$$H = -\text{grad } \Phi \tag{25}$$

If μ is constant, (24) reduces to Laplace's equation.

(vi) For flow of steady currents, the conduction current vector $\vec{j}$ may be derived from a potential function Φ through

$$\vec{j} = -\sigma \text{ grad } \Phi, \tag{26}$$

where σ is the conductivity, then Φ satisfies the equation

$$\text{div } (\sigma \text{ grad } \Phi) = 0, \tag{27}$$

which reduces to Laplace's equation when σ is constant.

(vii) The velocity potential Φ of two dimensional wave motion of small amplitude in a perfect fluid under gravity satisfies Laplace's equation.

(viii) For steady flow in the theory of conduction of heat, the temperature T satisfies the equation

$$\text{div } (k\nabla T) = 0, \tag{28}$$

where k is the thermal conductivity. This reduces to Laplace's equation if k is constant.

6.2.6 Mathematical Modelling in Terms of Diffusion Equation

(i) In the absence of heat sources or sinks, the temperature T satisfies the diffusion equation

$$\frac{\partial T}{\partial t} = \frac{k}{\rho c} \nabla^2 T \tag{29}$$

(ii) If c is the concentration of a diffusing substance, then the diffusing current vector $\vec{J}$ is given by Fick's first law of diffusion in the form

$$\vec{J} = -D \operatorname{grad} c, \tag{30}$$

where D is the coefficient of diffusion for the substance under consideration.

The equation of continuity for the diffusion substance is deduced as in Section 6.2.1 as

$$\frac{\partial c}{\partial t} + \operatorname{div} \vec{J} = 0 \tag{31}$$

From (30) and (31)

$$\frac{\partial c}{\partial t} = \operatorname{div}(D \operatorname{grad} c), \tag{32}$$

which reduces to the diffusion equation if D is constant.

(iii) The vorticity vector $\vec{\zeta}$ which is defined as the curl of the velocity vector $\vec{V}$ of a fluid satisfies the equation

$$\frac{\partial \vec{\zeta}}{\partial t} = \frac{\mu}{\rho} \boldsymbol{\nabla}^2 \vec{\zeta} = \nu \boldsymbol{\nabla}^2 \vec{\zeta}, \tag{33}$$

when the motion is started from rest. Here μ is the coefficient of viscosity, ρ is the density and ν is the kinematic viscosity of the fluid.

(iv) For conducting media, Maxwell's equations of electromagnetism give

$$\boldsymbol{\nabla}^2 \vec{E} = \frac{K\mu}{c^2} \frac{\partial^2 \vec{E}}{\partial t^2} + \frac{4\pi\sigma\mu}{c^2} \frac{\partial \vec{E}}{\partial t} \tag{34}$$

where σ is conductivity, μ is permeability, K is dielectric constant. For propagation of long waves in a good conductor, the first term on the RHS can be neglected in comparison with the third and (37) reduces to the diffusion equation.

(v) When there is no production of neutrons, the one-dimensional transport equation governing the slowing down of neutrons in matter can be written in the form

$$\frac{\partial N}{\partial \theta} = \frac{\partial^2 N}{\partial z^2}, \tag{35}$$

where $N(z, \theta)$ is the number of neutrons per unit time which reach the age θ.

(iv) With species diffusion in space, Volterra's equations for n interacting species are modified to

$$\frac{\partial N_i}{\partial t} = k_i N_i + N_i \beta_i \sum_{j=1}^{N} a_{ij} N_j + D_i \left(\frac{\partial^2 N_i}{\partial x^2} + \frac{\partial^2 N_i}{\partial y^2} + \frac{\partial^2 N_i}{\partial z^2} \right) \tag{36}$$

$$i = 1, 2, \ldots, n$$

For diffusion of one species in one-dimensional space, it becomes

$$\frac{\partial N}{\partial t} = aN - bN^2 + D \frac{\partial^2 N}{\partial x^2} \tag{37}$$

EXERCISE 6.2

1. Consider a volume element in the shape of a rectangular parallelopiped, centred at the point x, y, z and with edges of length $\Delta x, \Delta y, \Delta z$ parallel to the axes of coordinates. Show that the mass of the fluid leaving the two faces perpendicular to x-axis, per unit time is given by

$$\frac{\partial}{\partial x}(\rho u)\, \Delta x\, \Delta y\, \Delta z \tag{38}$$

Use this result to deduce (6), without using Gauss's Divergence Theorem.

2. By using the method of Ex. 1 and suitable volume elements, show that the equation of continuity in spherical polar coordinates is

$$\frac{\partial \rho}{\partial t} + \frac{1}{r^2} \frac{\partial}{\partial r}(\rho q_r r^2) + \frac{1}{r \sin \theta}(\rho q_\theta \sin \theta) + \frac{1}{r \sin \theta} \frac{\partial}{\partial \Phi}(\partial q_\phi) = 0 \tag{39}$$

3. By using the method of Ex. 2, show that the equation of continuity in cylindrical polar coordinates is

$$\frac{\partial \rho}{\partial t} + \frac{1}{r} \frac{\partial}{\partial r}(\rho v_r r) + \frac{1}{r} \frac{\partial}{\partial \theta}(\rho v_\theta) + \frac{\partial}{\partial}(\rho v_z) = 0 \tag{40}$$

4. Show that in spherical polar coordinates

$$\nabla^2 \Psi \equiv \frac{1}{r^2} \frac{\partial}{\partial r}\left(r^2 \frac{\partial \Psi}{\partial r}\right) + \frac{1}{r^2 \sin \theta} \frac{\partial}{\partial \theta}\left(\sin \theta \frac{\partial \Psi}{\partial \theta}\right) + \frac{1}{r^2 \sin \theta} \frac{\partial^2 \Psi}{\partial \Phi^2} \tag{41}$$

5. Show that in cylindrical polar coordinates

$$\nabla^2 \Psi \equiv \frac{1}{r} \frac{\partial}{\partial r}\left(r \frac{\partial \Psi}{\partial r}\right) + \frac{1}{r^2} \frac{\partial^2 \Psi}{\partial \theta^2} + \frac{\partial^2 \Psi}{\partial z^2} \tag{42}$$

6. Write the diffusion equation in spherical polar and cylindrical polar coordinates.

7. Show that a general solution of Laplace's equation which is independent of Φ has the form

$$\sum_n \left(A_n r^n + \frac{B_n}{r^{n+1}}\right) P_n (\cos \theta), \tag{43}$$

where $P_n(\mu)$ satisfies Lagendres equation

$$(1 - \mu^2) \frac{d^2 P_n}{d\mu^2} - 2\mu \frac{dP_n}{d\mu} + n(n + 1)P_n = 0 \tag{44}$$

8. Show that the diffusion equation

$$\frac{\partial^2 \theta}{\partial x^2} = \frac{1}{K} \frac{\partial \theta}{\partial t} \tag{45}$$

is satisfied by

(i) $\quad \theta = \dfrac{1}{\sqrt{t}} \exp\left(-\dfrac{x^2}{4Kt}\right)$ (46)

(ii) $\quad \theta = \dfrac{1}{2\sqrt{\pi K}} \exp\left[-\dfrac{(x-\xi)^2}{4Kt}\right)$ (47)

(iii) $\quad \theta = \dfrac{1}{2\sqrt{\pi Kt}} \displaystyle\int_{-\infty}^{\infty} \Phi(\xi) \exp\left[-\dfrac{(x-\xi)^2}{4Kt}\right] d\xi,$ (48)

where in (47), ξ is arbitrary constant and in (48) $\Phi(\xi)$ is an arbitrary continuous function of ξ.

9. Deduce equation (15) from first principles. State all the assumptions underlying its derivation explicitly.

10. Attempt the mathematical derivation of all the thirteen models given in Sections 6.2.5 and 6.2.6.

6.3 MOMENTUM-BALANCE EQUATIONS: THE SECOND METHOD OF OBTAINING PARTIAL DIFFERENTIAL EQUATION MODELS

6.3.1 Euler's Equations of Motion for Inviscid Fluid Flow

In an inviscid fluid, the force due to the fluid on any immersed plane area is always normal to it. Accordingly the forces on any fluid element due to rest of the fluid are always normal to the bounding surface at every point. Thus the resultant force due to the rest of the fluid on the given element (Fig. 6.2)

$$= -\iint_S p\,d\vec{S} = -\iint_S p\hat{n}\,dS = -\iiint_T \operatorname{grad} p \, dx\, dy\, dz,$$ (49)

so that the equation of motion for this fluid element is

$$\iiint_T \rho\frac{d\vec{V}}{dt} dx\, dy\, dz = \iiint_T \rho\vec{F}\, dx\, dy\, dz - \iiint_T \nabla p\, dx\, dy\, dz$$ (50)

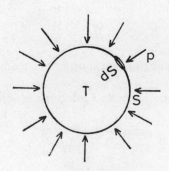

Figure 6.2

where $\vec{F}$ is the external body force per unit mass. This gives

$$\rho \frac{d\vec{V}}{dt} = \rho \vec{F} - \boldsymbol{\nabla} p \tag{51}$$

or

$$\frac{\partial u}{\partial t} + u \frac{\partial u}{\partial x} + v \frac{\partial u}{\partial y} + w \frac{\partial u}{\partial z} = F_x - \frac{1}{\rho} \frac{\partial p}{\partial x}$$

$$\frac{\partial v}{\partial t} + u \frac{\partial v}{\partial x} + v \frac{\partial v}{\partial y} + w \frac{\partial v}{\partial z} = F_y - \frac{1}{\rho} \frac{\partial p}{\partial y} \tag{52}$$

$$\frac{\partial w}{\partial t} + u \frac{\partial w}{\partial x} + v \frac{\partial w}{\partial y} + w \frac{\partial w}{\partial z} = F_z - \frac{1}{\rho} \frac{\partial p}{\partial z}$$

Equation (6) and (52) give us four coupled equations to determine $u(x, y, z, t)$, $v(x, y, z, t)$, $w(x, y, z, t)$ and $p(x, y, z, t)$. For a compressible fluid, ρ is variable and we need a fifth equation which is given by the equation of state

$$p = f(\rho) \tag{53}$$

For compressible inviscid fluids, (6), (52) and (53) give us five equations to determine u, v, w, p and ρ.

For viscous fluids, in addition to normal pressure forces, there are tangential viscous forces also and as such equations (52) have to be modified. For Newtonian viscous fluids for which the relation between stress and strain-rate tensor is linear and for which the viscosity coefficient μ is constant, the modification consists of addition of terms $\mu \nabla^2 u$, $\mu \nabla^2 v$, $\mu \nabla^2 w$ to the right hand sides of the three equations of (52). For Non-Newtonian fluids for which the relation between stress and strain-rate tensors is nonlinear, the modifications are much more complicated.

Moreover due to viscous dissipation, heat may be generated, temperature may change and to determine this new variable, an additional equation is necessary. This is given by the energy equation.

6.3.2 Partial Differential Equation Model for a Vibrating String

Let T be the tension of the elastic string held tightly between the points A and B corresponding to $x = 0$ and $x = L$. Let the string be slightly disturbed. Let $u(x, t)$ be the displacement at time t of an element of original length Δx and mass $\rho \Delta x$.

The force on this element in the direction of the displacement (Fig. 6.3)

$$= (T \sin \Psi)_{x+\Delta x} - (T \sin \Psi)_x$$
$$= f(x + \Delta x) - f(x); \qquad f(x) = T \sin \Psi$$
$$\simeq \Delta x f'(x) = \Delta x \frac{\partial}{\partial x}(T \sin \Psi)$$
$$\simeq \Delta x \frac{\partial}{\partial x}(T \tan \Psi) = \Delta x \frac{\partial}{\partial x}\left(T \frac{\partial u}{\partial x}\right) = \Delta x T \frac{\partial^2 u}{\partial x^2}, \tag{54}$$

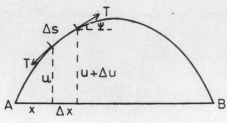

Figure 6.3

so that the equation of motion for this element is

$$\rho \Delta x \frac{\partial^2 u}{\partial t^2} = T \frac{\partial^2 u}{\partial x^2} \Delta x$$

or

$$\frac{\partial^2 u}{\partial x^2} = \frac{\rho}{T} \frac{\partial^2 u}{\partial t^2} = \frac{1}{c^2} \frac{\partial^2 u}{\partial t^2}; \quad c^2 = \frac{T}{\rho} \tag{55}$$

This is the *wave equation* in one dimension.

6.3.3 Partial Differential Equation Model for a Vibrating Membrane

Here let $u(x, y, t)$ be the displacement at time t of an element of original area $\Delta x \, \Delta y$ and mass $\rho \, \Delta x \, \Delta y$, then proceeding as in Section 6.3.2, we get

$$\rho \, \Delta x \, \Delta y \frac{\partial^2 u}{\partial t^2} = T \Delta y \frac{\partial}{\partial x}\left(\frac{\partial T}{\partial x}\right)\Delta x + \Delta x T \frac{\partial}{\partial y}\left(\frac{\partial u}{\partial y}\right) \Delta y$$

or

$$\frac{\partial^2 u}{\partial t^2} = \frac{T}{\rho}\left(\frac{\partial^2 u}{\partial x^2} + \frac{\partial^2 u}{\partial y^2}\right) \quad \text{or} \quad \frac{\partial^2 u}{\partial x^2} + \frac{\partial^2 u}{\partial y^2} = \frac{1}{c^2} \frac{\partial^2 u}{\partial t^2} \tag{56}$$

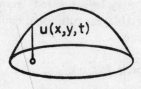

This is the wave equation in two-dimensions. Similarly the wave equation in three dimensions is

$$\frac{\partial^2 u}{\partial x^2} + \frac{\partial^2 u}{\partial y^2} + \frac{\partial^2 u}{\partial z^2} = \frac{1}{c^2} \frac{\partial^2 u}{\partial t^2}$$

$$\tag{57}$$

Figure 6.4

6.3.4 Mathematical Modelling in Terms of Wave Equation

(i) *Transverse vibrations of a string.* This has been discussed in Section 6.3.2.

(ii) *Transverse Vibrations of a membrane.* This has been discussed in Section 6.3.3.

(iii) *Longitudinal Vibrations in a bar:* If a uniform bar of elastic material of uniform cross section placed along the x-axis is stressed in such a way that each point of a typical cross-section has the same displacement $u(x, t)$, then $u(x, t)$ satisfies (55) where $c^2 = E/\rho$, E being the Young's modulus and ρ the density of material of the bar.

(iv) *Longitudinal Sound Waves:*   If plane waves of sound are propagated in a cylindrical vessel whose cross-sectional area at x is $A(x)$ in such a way that every point of this cross-section has the same longitudinal displacement $u(x, t)$, then we have

$$\frac{\partial}{\partial x}\left[\frac{1}{A}\frac{\partial}{\partial x}(Ax)\right] = \frac{1}{c^2}\frac{\partial^2 u}{\partial t}, \tag{58}$$

which reduces to (55) when A is constant and c denotes the local velocity of sound.

(v) *Sound Waves in Space:*   Let the pressure and density of a gas change from p_0, ρ_0 to p, ρ due to the passage of sound wave in it, so that for a small disturbance

$$\rho = \rho_0(1 + s), \qquad p = p_0 + c^2\rho_0 s \tag{59}$$

For small disturbances, the equations of motion and continuity give

$$\rho_0 \frac{\partial \vec{V}}{\partial t} = -c^2\rho_0 \boldsymbol{\nabla}s, \qquad \rho_0 \frac{\partial s}{\partial t} + \rho_0 \operatorname{div} \vec{V} = 0 \tag{60}$$

If the motion is irrotational $\vec{V} = -\operatorname{grad} \Phi$ and equation (60) give

$$\operatorname{grad}\left(\frac{\partial \Phi}{\partial t} - c^2 s\right) = 0, \quad \frac{\partial s}{\partial t} = \boldsymbol{\nabla}^2\Phi \tag{61}$$

Eliminating s between these, we get the wave equation

$$\boldsymbol{\nabla}^2\Phi = \frac{1}{c^2}\frac{\partial^2\Phi}{\partial t^2} \tag{62}$$

(vi) *Electromagnetic Waves:*   If we define A and Φ by

$$\vec{H} = \operatorname{curl} \vec{A}, \; \vec{E} = -\frac{1}{c}\frac{\partial R}{\partial t} - \boldsymbol{\nabla}\Phi, \tag{63}$$

then Maxwell's equations of electromagnetic theory viz.

$$\operatorname{div} \vec{E} = 4\pi\rho, \quad \operatorname{div} \vec{H} = 0, \quad \operatorname{curl} \vec{E} = -\frac{1}{c}\frac{\partial \vec{H}}{\partial t},$$

$$\operatorname{curl} \vec{H} = \frac{4\pi i}{c} + \frac{1}{c}\frac{\partial \vec{E}}{\partial t} \tag{64}$$

are satisfied if

$$\boldsymbol{\nabla}^2\vec{A} = \frac{1}{c^2}\frac{\partial^2\vec{A}}{\partial t^2} - \frac{4\pi}{c}\vec{i}, \; \boldsymbol{\nabla}^2\Phi = \frac{1}{c^2}\frac{\partial^2\Phi}{\partial t^2} - 4\pi\rho, \tag{65}$$

so that in the absence of charges or currents, Φ and the components of vector $\vec{A}$ satisfy the wave equation.

(vii) *Elastic Waves in Solids:*   If the displacement vector $\vec{V}$ is written as

$$\vec{V} = \operatorname{grad} \Phi + \operatorname{curl} \vec{\Psi}, \tag{66}$$

then it can be shown that in the absence of body forces, Φ, $\vec{\Psi}$ satisfy the wave equation

$$\frac{\partial^2 \Phi}{\partial t^2} = c_1^2 \nabla^2 \Phi, \quad \frac{\partial^2 \vec{\Psi}}{\partial t^2} = c_2^2 \nabla^2 \vec{\Psi}, \tag{67}$$

where

$$c_1^2 = \frac{\lambda + 2\mu}{\rho}, \quad c_2^2 = \frac{\mu}{\rho}, \tag{68}$$

and λ, μ are Lame's constants.

EXERCISE 6.3

1. Show that the equations (52), (55), (56), (57), (58), (62), (65) are dimensionally correct.

2. Show that $u = f(x + ct) + g(x - ct)$, where $f(\cdot)$ and $g(\cdot)$ are arbitrary continuous functions, satisfies (55).

3. Show that if $u(x - t) = g(x - ct)$, then $u(x + kc, \overline{t} + k) = u(x, t)$. Interpret the solution as a wave propagating forward with velocity c. Similarly interpret the solution $u = f(x + ct)$ as a wave propagating backward with velocity c.

4. Show that an appropriate solution of (57) which vanishes at $x = 0$, $x = a$, $y = 0$, $y = b$ is given by

$$u(x, y, t) = \sum_{m,n} A_{m,n} \sin\left(\frac{m\pi x}{a}\right) \sin\left(\frac{n\pi y}{b}\right) \cos(k_{mn}ct), \tag{69}$$

where

$$k_{mn}^2 = \pi^2 \left(\frac{m^2}{a^2} + \frac{n^2}{b^2}\right) \tag{70}$$

5. Attempt to derive the seven mathematical models of Section 6.3.4.

6.4 VARIATIONAL PRINCIPLES: THIRD METHOD OF OBTAINING PARTIAL DIFFERENTIAL EQUATION MODELS

6.4.1 Euler-Lagrange Equation

Let

$$I = \iint_S F(x, y, u, u_x, u_y) \, dx \, dy \tag{71}$$

where $F(\ )$ is a known function, then the value of I depends on $u(x, y)$ and our object is to choose $u(x, y)$ so that the integral I has a maximum or minimum value. Such a function is given by Euler-Lagrange equation of calculus of variations *viz.*

$$\frac{\partial F}{\partial u} - \frac{\partial}{\partial x}\left(\frac{\partial F}{\partial u_x}\right) - \frac{\partial}{\partial y}\left(\frac{\partial F}{\partial u_y}\right) = 0 \tag{72}$$

Since F is a known function of x, y, u, u_x, u_y, therefore $\partial F/\partial u$, $\partial F/\partial u_x$, $\partial F/\partial u_y$ are also known functions of x, y, u, u_x, u_y. As such the left hand side of (72) is a known function of x, y, u, u_x, u_y, u_{xx}, u_{xy}, u_{yy} so that (72) gives a partial differential equation of second order for determining $u(x, y)$.

6.4.2 Minimal Surfaces

To illustrate the use of (72), we consider the problem of finding the surface with minimum area out of all those surfaces which are bounded by a given skew curve. The surface area is given by

$$I = \iint_S \sqrt{1 + p^2 + q^2}\, dx\, dy; \quad p = \frac{\partial z}{\partial x}, \quad q = \frac{\partial z}{\partial y} \tag{73}$$

so that (72) gives

$$0 - \frac{\partial}{\partial x}\left(\frac{p}{\sqrt{1 + p^2 + q^2}}\right) - \frac{\partial}{\partial y}\left(\frac{a}{\sqrt{1 + p^2 + q^2}}\right) = 0 \tag{74}$$

or

$$(1 + q^2)r + (1 + p^2)t - 2pqs = 0; \quad r = \frac{\partial^2 z}{\partial x^2}, \quad s = \frac{\partial^2 z}{\partial x \partial y}, \quad t = \frac{\partial^2 z}{\partial y^2} \tag{75}$$

Now if (75) is satisfied, then the sum of the principal radii of curvature at every point of the surface is zero i.e. the mean curvature is zero at every point. A surface for which the mean curvature is zero at every point is called a minimal surface and the above discussion explains the reason for this.

It can be shown that the only ruled surface which is a minimal surface is a right helicoid. It can also be shown that the catenoid obtained by rotating a catenary about its dirrectrix is a minimal surface.

6.4.3 Vibrating String

Here we apply Hamilton's principle according to which the shape of the string is to be such that

$$I = \int_0^{t_0} (T - V)\, dt, \tag{76}$$

is minimum where T is the kinetic energy and V is the potential energy of the string. Using the notation of Section 6.3.2,

$$T = \frac{1}{2}\rho \int_0^L \left(\frac{\partial u}{\partial t}\right)^2 dx \tag{77}$$

To obtain the potential energy, we find the work done in stretching the string from its natural length L to the present length so that

$$V = T\left[\int_0^L \sqrt{1 + (\partial u/\partial x)^2}\, dx - L\right]$$

$$\simeq \frac{1}{2}T \int_0^L (\partial u/\partial x)^2\, dx \tag{78}$$

From (76), (77) and (78)

$$I = \frac{1}{2} \int_0^L \int_0^{t_0} \left[\rho(\partial u/\partial t)^2 - T(\partial u/\partial x)^2 \right] dx\, dt \tag{79}$$

Using (72)

$$\frac{\partial}{\partial t}\left(\rho \frac{\partial u}{\partial t} \right) - \frac{\partial}{\partial x}\left(T \frac{\partial u}{\partial x} \right) = 0$$

or

$$\frac{\partial^2 u}{\partial x^2} = \frac{1}{c^2} \frac{\partial^2 u}{\partial t^2}, \quad c^2 = \frac{T}{\rho}, \tag{80}$$

which is the same as (55).

6.4.4 Vibrating Membrane

Here

$$T = \frac{1}{2} \iint_S \rho \left(\frac{\partial u}{\partial t} \right)^2 dx\, dy, \tag{81}$$

and the potential energy V is obtained by finding the work done in stretching the membrane from its original area to the new surface area so that

$$V = T \left[\iint_S \sqrt{1 + (\partial u/\partial x)^2 + (\partial u/\partial y)^2}\, dx\, dy - \iint_S dx\, dy \right]$$

$$\simeq \frac{1}{2} T \iint \left[(\partial u/\partial x)^2 + (\partial u/\partial y)^2 \right] dx\, dy \tag{82}$$

Then

$$I = \frac{1}{2} \iiint \{ \rho(\partial u/\partial t)^2 - [(\partial u/\partial x)^2 + (\partial u/\partial y)^2] \}\, dt\, dx\, dy \tag{83}$$

Using Euler-Lagrange equation of calculus of variation

$$\frac{\partial}{\partial t}\left(\rho \frac{\partial u}{\partial t} \right) - \frac{\partial}{\partial x}\left(T \frac{\partial u}{\partial x} \right) - \frac{\partial}{\partial y}\left(T \frac{\partial u}{\partial y} \right) = 0$$

or

$$\frac{\partial^2 u}{\partial x^2} + \frac{\partial^2 u}{\partial y^2} = \frac{1}{c^2} \frac{\partial^2 u}{\partial t^2}, \quad c^2 = \frac{T}{\rho}, \tag{85}$$

which is the same as (56).

6.4.5 Gas-Filled Cylinder

The mathematical discussion for the case of a vibrating string applies to the longitudinal vibrations of any elastic medium. In particular it applies to longitudinal vibrations of an elastic bar (6.3.4(iii)). It also applied to vibrations of a gas in a cylinder. Instead of ρ, we shall have the mass per unit volume and instead of T, we shall have a constant depending on the compressibility of the gas.

EXERCISE 6.4

1. Prove that the total energy of a string which is fixed at the points $x = 0$, $x = L$ and is executing small transverse vibrations is

$$W = \frac{1}{2}T\int_0^L \left[(\partial u/\partial x)^2 + \frac{1}{c^2}(\partial u/\partial t)^2 \right] dx \tag{87}$$

Show that if $u = f(x - ct)$ and $0 \le x \le L$, then the energy of the wave is equally divided between potential energy and kinetic energy. Does this result hold for

(i) $u = g(x + ct)$ (ii) $u = f(x - ct) + g(x + ct)$? $\tag{88}$

2. Discuss the problem corresponding to that of Ex. 1 for a rectangular vibrating membrane for the solution given by (69).

3. Show that $u = A(p) \exp [ip(t \pm x/c]$

is a solution of the one-dimensional wave-equation for arbitrary form of the function A which depends on p only. Interpret these solutions physically.

6.5 PROBABILITY GENERATING FUNCTION, FOURTH METHOD OF OBTAINING PARTIAL DIFFERENTIAL EQUATION MODELS

6.5.1 P.D.E. Model for Birth-death-immigration-emigration Process

Let $p(n, t)$ denote the probability of there being n persons in the population at time t. Also let $n\lambda \, \Delta t + o(\Delta t)$, $n\mu \, \Delta t + o(\Delta t)$, $\nu \Delta t + o(\Delta t)$, $\alpha \Delta t + o(\Delta t)$ denote respectively the probabilities of a single birth, a single death, a single immigration and a single emigration in the time interval $(t, t + \Delta t)$ and let the probability of more than one event in this time interval be $o(\Delta t)$, then by using the theorems of total and compound probabilities, we get

$$\begin{aligned} p(n, t + \Delta t) = &\ p(n + 1, t)((n + 1)\mu\Delta t + \alpha\Delta t + o(\Delta t)) \\ &+ p(n - 1, t)((n - 1)\lambda\Delta t + \nu\Delta t + o(\Delta t)) \\ &+ p(n, t)(1 - n\lambda\Delta t - n\mu\Delta t - \alpha\Delta t - \nu\Delta t \\ &- (o(\Delta t)) \qquad n = 1, 2, 3, \ldots \end{aligned} \tag{89}$$

$$\begin{aligned} p(0, t + \Delta t) = &\ p(1, t)(\mu\Delta t + \alpha\Delta t + o(\Delta t) \\ &+ p(0, t)(1 - \nu\Delta t - o(\Delta t) \end{aligned} \tag{90}$$

Transferring $p(n, t)$ and $p(0, t)$ to the left hand sides, dividing by Δt and proceeding to the limit as $\Delta t \to 0$, we obtain the following system of differential-difference equations for a BDIE process:

$$\begin{aligned} p'(n, t) = &\ [(n + 1)\mu + \alpha]p(n + 1, t) - [n(\mu + \lambda) + \alpha + \nu]p(n, t) \\ &+ [n - 1)\lambda + \nu]p(n - 1, t); \quad n = 1, 2, 3, \ldots \end{aligned} \tag{91}$$

$$p'(0, t) = (\mu + \alpha)p(1, t) - \nu p(0, t) \tag{92}$$

Defining the probability generating function

$$\Phi(s, t) = \sum_{n=0}^{\infty} p(n, t)s^n, \tag{93}$$

we get on multiplying (91) by s^n, (92) s^0 and summing for all values of n:

$$\frac{\partial \Phi}{\partial t} = (\lambda s - \mu)(s - 1)\frac{\partial \Phi}{\partial s} + \left(\nu - \frac{\alpha}{s}\right)(s - 1)\Phi + \frac{\alpha(s - 1)}{s}p(0, t) \quad (94)$$

If $\alpha = 0$, this is a linear partial equation of the first order. For $\nu = 0, \alpha = 0$, it has been solved in the literature. For $\nu = 0$, $\alpha \neq 0$, this equation has only recently been solved by Kapur. Once it is solved, $p(n, t)$ can be obtained for all values of n, by using (93).

6.5.2 P.D.E. Model for a Stochastic Epidemic Process with No Removal

Let $p_n(t)$ be the probability that there are n susceptible persons in the system and let $f_j(n)\Delta t + o(\Delta t)$ give the probability that the number will change to $n + j$ in the time-interval $(t, t + \Delta t)$. Here j is any positive or negative integer and $o(\Delta t)$ is an infinitesimal which is such that $o(\Delta t)/\Delta t \to 0$ as $\Delta t \to o$. The probability that there is no change in the time-interval $(t, t + \Delta t)$ is then given by $1 - \sum_{j \neq 0} f_j(n)\Delta t + o(\Delta t)$. Using the theorems of total and compound probabilities, we get

$$p_n(t + \Delta t) = p_n(t)(1 - \sum_{j \neq 0} f_j(n)\Delta t]) + \sum_{j \neq 0} p_{n-j}(t)f_j(n - j)\Delta t + o(\Delta t) \quad (95)$$

Transferring $p_n(t)$ to the left-hand-side, dividing by Δt and taking the limit as $\Delta t \to 0$, we get

$$\frac{dp_n}{dt} = - p_n(t) \sum_{j \neq 0} f_j(n) + \sum_{j \neq 0} p_{n-j}(t)f_j(n - j) \quad (96)$$

Multiplying (96) by s^n, summing for all n and using the definition of the probability generating function viz. (93), we get

$$\frac{\partial \Phi}{\partial t} = - \sum_{j \neq 0} \sum_n f_j(n)p_n s^n + \sum_{j \neq 0} \sum_n p_{n-j}(t)f_j(n - j)s^{n-j}, \quad (97)$$

giving the basic partial differential equation

$$\frac{\partial \Phi}{\partial t} = \sum_{j \neq 0} (s^j - 1)f_j\left(s\frac{\partial}{\partial s}\right)\Phi(s, t) \quad (98)$$

For a two-dimensional stochastic process, the corresponding partial differential equation obtained in the same manner is

$$\frac{\partial \Phi}{\partial t} = \sum_{j \neq 0} \sum_{k \neq 0} (u^j v^k - 1)f_{j,k}\left(u\frac{\partial}{\partial x}, v\frac{\partial}{\partial y}\right)\Phi(u, v, t), \quad (99)$$

where $\qquad \Phi(u, v, t) = \sum_m \sum_n p(m, n, t)u^m v^n \qquad (100)$

$p(m, n, t)$ is the probability of there being m individuals of the first kind and n individuals of the second kind and $f_{jk}(m, n)\Delta t + o(\Delta t)$ is the probability of the number of the two kinds changing from m to $m + j$ and n to $n + k$ in the time interval $(t, t + \Delta t)$.

6.5.3 Stochastic Epidemic Model with No Removal

Let there be initially at $t = 0$, n susceptibles and 1 infective in the system.

Also let the probability of there being r susceptible persons at time t be $p(r, t)$. We assume that the probability of one more person becoming infected in time Δt is $\beta r(n + 1 - r)\Delta t + o(\Delta t)$, so that

$$\left. \begin{array}{ll} f_j(r) = \beta r(n + 1 - r) & \text{when } j = -1 \\ \\ \qquad = 0 & \text{when } j \neq -1 \end{array} \right\} \tag{101}$$

Substituting (101) in (98), we get

$$\frac{\partial \Phi}{\partial t} = \beta(s^{-1} - 1) \left[s \frac{\partial}{\partial s} \left(n + 1 - s \frac{\partial}{\partial s} \right) \Phi \right]$$

$$= \beta(1 - s) \left[(n + 1) \frac{\partial \Phi}{\partial s} - \frac{\partial \Phi}{\partial s} - s \frac{\partial^2 \Phi}{\partial s^2} \right]$$

or

$$\frac{\partial \Phi}{\partial t} = \beta(1 - s) \left(n \frac{\partial \Phi}{\partial s} - s \frac{\partial^2 \Phi}{\partial s^2} \right) \tag{102}$$

EXERCISE 6.5

1. Substituting from (93) in (102) and equating coefficients of various powers of s, prove that

$$\frac{dp_r}{dt} = \beta(r + 1)(n - 1)p_{r-1} - \beta r(n + r + 1)p_r;$$

$$r = 0, 1, 2, \ldots, n - 1 \tag{103}$$

$$\frac{dp_0}{dt} = -\beta n p_0 \tag{104}$$

Also show that the initial conditions are

$$p_n(0) = 1, \quad p_r(0) = 0, \quad \text{when } r = 0, 1, 2, \ldots n - 1$$

2. Integrate (101) and (102) subject to (103) to show that

$$p_n(t) = \exp(-\beta t)$$

$$p_{n-1}(t) = \frac{n}{n - 2} (\exp(-n\beta t) - \exp(-2n - 2)\beta t) \tag{105}$$

6.6 MODEL FOR TRAFFIC ON A HIGHWAY

6.6.1 Relation Between Car Velocity u and Traffic Density ρ

For discussing traffic flow on a highway, we need, in addition to (15), another relation between u and ρ. We don't have here a simple momentum balance equation as we had in the case of fluid dynamics which enabled us to get Euler's equation (52). We can either obtain this relation empirically or derive it from other hypotheses. We consider the car-following model in which we assume that the acceleration of a car is proportional to the difference between the speed of the car ahead of the $(n - 1)$th car and the car itself (the nth car) so that (Figure 6.5)

$$\frac{d^2 x_n}{dt^2} = -\lambda \left(\frac{dx_n}{dt} - \frac{dx_{n-1}}{dt} \right) \tag{106}$$

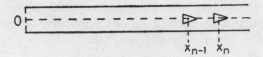

Figure 6.5

Integrating (106), wc get

$$\frac{dx_n}{dt} = \lambda(x_{n-1} - x_n) + d_n \tag{107}$$

so that

$$u = \frac{\lambda}{\rho} + d, \tag{108}$$

where ρ is the traffic density. We assume that at maximum traffic-density (bumper to bumper) ρ_{max}, the velocity u is zero so that

$$0 = \frac{\lambda}{\rho_{max}} + d, \tag{109}$$

From (108) and (109)

$$u = \lambda \left(\frac{1}{\rho} - \frac{1}{\rho_{max}} \right) \tag{110}$$

This will imply (Figure 6.6) that $u \to \infty$ as $\rho \to 0$ i.e. when the road is empty except for one car. Usually there is a speed limitation on every highway. Let this speed limit be u_{max}, then

$$\left. \begin{array}{ll} u = u_{max} & \text{when} \quad \rho \leq \rho_1 \\ u = \left(\dfrac{1}{\rho} - \dfrac{1}{\rho_{max}} \right), & \rho \geq \rho_1 \end{array} \right\} \tag{111}$$

where

$$u_{max} = \lambda \left(\frac{1}{\rho_1} - \frac{1}{\rho_{max}} \right) \tag{112}$$

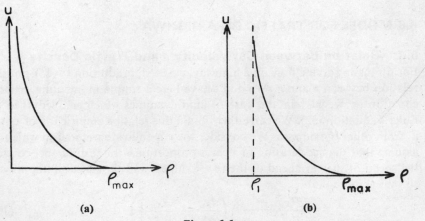

Figure 6.6

Here λ is the determined empirically, then (113) determines ρ_1 in terms of u_{max} and finally (111) gives the desired relation between u and ρ

6.6.2 An Alternative Relation Between u and ρ

For this derivation, we also assume that a driver's acceleration or deceleration also depends on the distance from the preceding car. The closer the driver is, the more strongly he is likely to respond to the observed car, so that we assume

$$\lambda = \frac{k}{x_{n-1}(t) - x_n(t)} \tag{113}$$

so that

$$\frac{d^2 x_n}{dt^2} = k \frac{\dfrac{dx_{n-1}}{dt} - \dfrac{dx_n}{dt}}{x_{n-1}(t) - x_n(t)} \tag{114}$$

Integrating (114), we get

$$u = -k \ln \rho + e \tag{115}$$

Also

$$0 = -k \ln \rho_{max} + e \tag{116}$$

so that

$$u = k \ln \frac{\rho_{max}}{\rho} \tag{117}$$

Again this implies that $u \to \infty$ as $\rho \to 0$. We modify this as follows

$$\left. \begin{array}{ll} u = u_{max} & \text{when} \quad \rho \leq \rho_2 \\ u = k \ln \dfrac{\rho_{max}}{\rho} & \text{when} \quad \rho \geq \rho_2 \end{array} \right\} \tag{118}$$

where

$$u_{max} = k \ln \frac{\rho_{max}}{\rho_2} \tag{119}$$

6.6.3 Traffic Wave Propagation Along a Highway

Since

$$\frac{\partial \rho}{\partial t} + \frac{\partial}{\partial x}(q) = 0; \qquad q = \rho u \tag{120}$$

and for small perturbations, we have

$$\rho = \rho_0 + \epsilon \rho_1(x, t), \qquad q = q_0 + \epsilon q_1(x, t) \tag{121}$$

We get

$$\frac{\partial \rho_1}{\partial t} + \frac{\partial q_1}{\partial x} = 0 \quad \text{or} \quad \frac{\partial \rho_1}{\partial t} + \left(\frac{dq}{d\rho}\right)_0 \frac{\partial \rho_1}{\partial x} = 0 \tag{122}$$

or

$$\frac{\partial \rho_1}{\partial t} + c \frac{\partial \rho_1}{\partial x} = 0; \quad c = (dq/d\rho)_0 \tag{123}$$

Equation (123) has the solution

$$\rho_1 = f(x - ct) \tag{124}$$

so that the disturbance propagates as a wave with velocity c. It propagates forward if $c > 0$ i.e. if $(dq/d\rho)_0 > 0$ or if the traffic is light (Figure 6.7) i.e. if the traffic density is less than a certain critical traffic density ρ_{crit} given by $dq/d\rho = 0$. Similarly the disturbance propagates backwards as a wave if the traffic is heavy and $\rho > \rho_{crit}$.

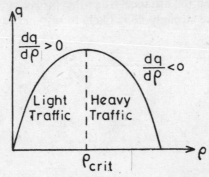

Figure 6.7

If u is given by (111), then

$$q = \rho u = \lambda\left(1 - \frac{\rho}{\rho_{max}}\right), \quad \frac{dq}{d\rho} = -\frac{\lambda}{\rho_{max}} < 0 \tag{125}$$

and the propagating wave always moves backward. On the other hand if u is given by (118), then

$$q = u\rho = k\rho \ln\frac{\rho_{max}}{\rho}, \quad \frac{dq}{d\rho} = k\left[\ln\frac{\rho_{max}}{\rho} - 1\right] \tag{126}$$

In this case the perturbation will propagate forward if $\rho < \rho < \rho_{max}/c$ and will otherwise propagate backwards.

The perturbation is constant along the straight lines

$$x - ct = \text{constant} \tag{127}$$

These are characteristics for the partial differential equation (123).

EXERCISE 6.6

1. Draw the diagrams of q against ρ for both the laws (111) and (118).

2. Discuss whether the velocity of propagation can be equal to the velocity of a car.

3. Integrate

$$\frac{d^2x_n}{dt^2} = k\frac{\dfrac{dx_{n-1}}{dt} - \dfrac{dx_n}{dx}}{[x_{n-1}(t) - x_n(t)]^a} \tag{128}$$

Deduce (111) and (118) when $a = 0$ and $a = 1$ respectively.

4. For Ex. 3, discuss the variation of ρ_{crit} with a.

6.7 NATURE OF PARTIAL DIFFERENTIAL EQUATIONS

6.7.1 Elliptic, Parabolic and Hyperbolic Equations

With the linear partial differential equation of the second order

$$\sum_{j=1}^{n} \sum_{i=1}^{n} a_{ij} \frac{\partial^2 \Phi}{\partial x_i \, \partial x_j} + \sum_{i=1}^{n} b_i \frac{\partial \Phi}{\partial x_i} + c\Phi = d, \tag{129}$$

We associate the characteristic hypersurface

$$\sum_{j=1}^{n} \sum_{i=1}^{n} a_{ij} x_i x_j + \sum_{i=1}^{n} b_i x_i = \text{constant} \tag{130}$$

The equation (129) is called elliptic, or hyperbolic according as the quadratic form $\sum_{j=1}^{n} \sum_{i=1}^{n} a_{ij} x_i x_j$ is positive definite, or indefinite. It is called parabolic if the determinant $| \, a_{ij} \, |$ vanishes. The reason is obvious since for the case of two independent variables, the corresponding curves are ellipses, hyperbolas and parabolas.

The distinction is important since characteristic curves or surfaces can be used in the solution of these equations.

6.7.2 Nature of Three Basic Linear Partial Differential Equations

The three basic linear partial differential equations of physics are:

The Laplace equation:
$$\frac{\partial^2 \Phi}{\partial x^2} + \frac{\partial^2 \Phi}{\partial y^2} + \frac{\partial^2 \Phi}{\partial z^2} = 0 \tag{131}$$

The Diffusion Equation:
$$\frac{\partial^2 \Phi}{\partial x^2} + \frac{\partial^2 \Phi}{\partial y^2} - k \frac{\partial \Phi}{\partial t} = 0 \tag{132}$$

The Wave Equation:
$$\frac{\partial^2 \Phi}{\partial x^2} + \frac{\partial^2 \Phi}{\partial y^2} - c^2 \frac{\partial^2 \Phi}{\partial t^2} = 0 \tag{133}$$

The matrices of the corresponding quadratic forms are

$$\begin{bmatrix} 1 & 0 & 0 \\ 0 & 1 & 0 \\ 0 & 0 & 1 \end{bmatrix}, \quad \begin{bmatrix} 1 & 0 & 0 \\ 0 & 1 & 0 \\ 0 & 0 & 0 \end{bmatrix} \begin{bmatrix} 1 & 0 & 0 \\ 0 & 1 & 0 \\ 0 & 0 & -c^2 \end{bmatrix} \tag{134}$$

so that the Laplace equation is elliptic, the diffusion equation is parabolic and the wave equation is hyperbolic.

Laplace equation usually arises in static or equilibrium situations e.g. in electrostatics, magnetostatics, gravitation, steady heat flow, flows of steady currents, irrotational fluid motion etc. (Section 6.2.5).

Diffusion equation arises when heat or population of a species or vorticity diffuses or mixes (Section 6.2.6).

Wave equation arises when disturbances propagate as in transverse vibrations of a string or a membrane or as sound waves or as light waves or as traffic waves (Section 6.3.3 and Section 6.6.3).

6.7.3 The Nature of the Partial Differential Equation for the Potential of the Steady Two-Dimensional Flow of Inviscid Flow of an Ideal Gas

The basic equations are

$$\frac{\partial u}{\partial x} + \frac{\partial v}{\partial y} + \frac{1}{\rho}\left(u\frac{\partial \rho}{\partial x} + v\frac{\partial \rho}{\partial y}\right) = 0 \tag{135}$$

$$u\frac{\partial u}{\partial x} + v\frac{\theta u}{\partial y} = -\frac{1}{\rho}\frac{\partial p}{\partial x}; \quad u\frac{\partial v}{\partial x} + v\frac{\partial v}{\partial y} = -\frac{1}{\rho}\frac{\partial p}{\partial y} \tag{136}$$

$$u = -\frac{\partial \Phi}{\partial x}, \quad v = -\frac{\partial \Phi}{\partial y}, \quad \frac{dp}{d\rho} = c^2, \tag{137}$$

where c is the local velocity of sound. From (135), (136), (137), we get

$$c^2\left(\frac{\partial u}{\partial x} + \frac{\partial v}{\partial y}\right) - u\left(u\frac{\partial u}{\partial x} + v\frac{\partial u}{\partial y}\right) - v\left(u\frac{\partial v}{\partial x} + v\frac{\partial v}{\partial y}\right) = 0 \tag{138}$$

In terms of the potential function, (138) gives

$$(c^2 - \Phi_x^2)\Phi_{xx} + (c^2 - \Phi_y^2)\Phi_{yy} - 2\Phi_x\Phi_y\Phi_{xy} = 0 \tag{139}$$

This equation is non-linear; in fact it is quasi-linear since it is linear in the second order derivatives only. The corresponding nature matrix is

$$\begin{bmatrix} c^2 - \Phi_x^2 & -\Phi_x\Phi_y \\ -\Phi_x\Phi_y & c^2 - \Phi_y^2 \end{bmatrix} \tag{140}$$

Thus the equation is elliptic, parabolic or hyperbolic according as

$$(c^2 - \Phi_x^2)(c^2 - \Phi_y^2) - \Phi_x^2\Phi_y^2 \lesseqgtr 0$$

or $$c^2 - (\Phi_x^2 + \Phi_y^2) \lesseqgtr 0 \quad \text{or} \quad q^2 \lesseqgtr c^2 \tag{141}$$

where q is the velocity of the fluid. Thus the potential equation (139) is elliptic, parabolic or hyperbolic according as the motion is subsonic, sonic or supersonic.

EXERCISE 6.7

1. Classify the following equations

   (i) $u_{xx} + u_{yy} + u_{zz} + u_{tt} = 0$

   (ii) $u_{xx} + u_{yy} + u_{zz} = ku_t$

   (iii) $u_{xx} + u_{yy} + u_{zz} = c^2 u_{tt}$

   (iv) $(1 + q^2)r + (1 + p^2)t - 2pqs = 0$

   (v) $u_{xx} + u_{yy} + u_{zz} = 0$.

2. Find the limiting form of (139) as $c \to \infty$. Interpret the result.

3. Show that the p.d.e.

$$R(x, y) \frac{\partial^2 z}{\partial x^2} + S(x, y) \frac{\partial^2 z}{\partial x \, \partial y} + T(x, y) \frac{\partial^2 z}{\partial y^2} + f\left(x, y, z, \frac{\partial z}{\partial x}, \frac{\partial z}{\partial y}\right) = 0$$

(142)

can be transformed to the forms

$$\frac{\partial^2 \zeta}{\partial \xi \, \partial \eta} = \Phi(\xi, \eta, \zeta, \zeta_\xi, \zeta_\eta) \qquad \text{where } S^2 - 4RT > 0 \qquad (143)$$

$$\frac{\partial^2 \zeta}{\partial \eta^2} = \Phi(\xi, \eta, \zeta, \zeta_\xi, \zeta_\eta) \qquad \text{when } S^2 - 4RT = 0 \qquad (144)$$

$$\frac{\partial^2 \zeta}{\partial \xi^2} + \frac{\partial^2 \zeta}{\partial \eta^2} = \Phi(\xi, \eta, \zeta, \zeta_\xi, \zeta_\eta) \qquad \text{where } S^2 - 4RT < 0 \qquad (145)$$

by suitable variate transformations

4. Transform $z_{xx} - x^2 z_{yy} = 0$ by the substitution

$$\xi \equiv y + \frac{1}{2} x^2, \qquad \eta \equiv y - \frac{1}{2} x^2.$$

5. Transform $z_{xx} + 2z_{xy} + z_{yy} = 0$ by the substitution

$$\xi = x - y, \qquad \eta = x + y$$

6. Transform $z_{xx} + x^2 z_{yy} = 0$ by the substitution $\xi = \frac{1}{2} x^2, \eta = y$.

6.8 INITIAL AND BOUNDARY CONDITIONS

Mathematical Modelling through Partial Differential Equations involves three essential steps:

(a) Formulation of the P.D.E.
(b) Specification of initial and boundary conditions.
(c) Solution of the P.D.E. subject to given initial and boundary conditions.

In the preceding sections we have mainly discussed step (a). We shall now briefly discuss step (b). Step (c) will not be discussed at all.

Laplace Equation
(i) *Gravitation*: When there is matter distributed over a surface, the potential function Φ can assume different forms Φ_1, Φ_2 on opposite sides of the surface, and on the surface these two functions satisfy the condition

$$\Phi_1 = \Phi_2, \qquad \frac{\partial \Phi_2}{\partial n} - \frac{\partial \Phi_1}{\partial n} = -4\pi\sigma, \qquad (146)$$

where σ is the surface density of matter and n is the normal to the surface directed from region 1 to region 2.

(ii) *Irrotational Motion of a perfect fluid*: If the fluid is at rest at infinity then $\Phi \to 0$ but if there is a velocity V in the z direction, then $\Phi \sim -Vz$ as

$z \to \infty$. When the fluid is in contact with a rigid surface which is moving so that a typical point of it moves with a velocity $\vec{U}$, then $(\vec{q} - \vec{U}) \cdot \vec{n} = 0$ where $\vec{n}$ is in the direction of the normal at P. The condition satisfied by Φ therefore is

$$\partial\Phi/\partial n = -(\vec{U} \cdot \vec{n}) \tag{147}$$

at all points on the surface. Also Φ should have no singularities except at sources on sinks.

(iii) *Electrostatics*: The potential function Φ is constant on any conductor Also at each point of a conductor

$$\partial\Phi/\partial n = -4\pi\sigma, \tag{148}$$

where σ is the surface density of the electric charge on the conductor. Also with a finite system of charges, $\Phi \to 0$ at infinity but if there is a uniform field E in the z direction, the $\Phi \sim -Ez$ as $z \to \infty$. Φ should have no singularities except at isolated charges, dipoles etc. If dielectric is present (148) is replaced by

$$K\partial\Phi/\partial n = -4\pi\sigma \tag{149}$$

and on the interface of two dielectrics

$$\Phi_1 = \Phi_2, \quad K_1 \frac{\partial\Phi_1}{\partial n} = K_2 \frac{\partial\Phi_2}{\partial n} \tag{150}$$

(iv) *Magnetostatics*: At a sudden changes of medium

$$\Phi_1 = \Phi_2, \quad \mu_1 \frac{\partial\Phi_1}{\partial n} = \mu_2 \frac{\partial\Phi_2}{\partial n} \tag{151}$$

Also in the presence of a constant field H_0 in the z-direction at infinity, we have $\Phi \sim -H_0 z$ as $z \to \infty$.

(v) *Steady Flow of Heat*: In this case $\partial T/\partial x = 0$ if there is no flux of heat across the boundary and $\partial T/\partial x + h(T - T_0) = 0$ where h is a constant when there is radiation from the surface into a medium at constant temperature T_0.

Heat Conduction Equation
Here the boundary conditions may specify the temperature at all points of the boundary surface, or may prescribe $\partial T/\partial n$ at all points of the surface or may require $\partial T/\partial n + h(T - T_0)$ to vanish when there is radiation from the surface into a medium of fixed temperature T_0.

Wave Equation
For vibration of a string, the initial condition specifies $u(x, 0)$. If the ends of the strings are fixed $u(0, t) = 0$, $u(L, t) = 0$ for all t. For vibrations of a membrane again $u(x, y, 0)$ has to be specified as a known function of x, y and if the edges of the membrane are fixed, then $u(x, y, t) = 0$ at all points

on the boundary for all values of t. For longitudinal vibrations of a bar we have

$$\frac{\partial u}{\partial t} = v(t) \quad \text{at } x = 0, \quad \text{where } v(t) \text{ is prescribed velocity of the end}$$

$$x = 0$$

$$\frac{\partial u}{\partial x} = 0 \quad \text{at } x = L, \quad \text{if end } x = L \text{ is free}$$

$$u = \partial u/\partial t = 0 \quad \text{at} \quad t = 0 \quad \text{for} \quad 0 \leq x \leq L.$$

For vibrations of gas in a cylinder with both ends open, we require

$$\frac{\partial u}{\partial x} = 0 \quad \text{at} \quad x = 0, \quad x = L \quad \text{for all } t.$$

Viscous Fluid Motion.   At any rigid boundary, the velocity of the fluid at any point of the boundary is the same as the given velocity of the boundary. At any porous boundary, the velocity of the fluid is determined by the suction or injection there.

Motion of a Conducting Fluid in a Magnetic Field

The boundary conditions on the velocity vector are simple. For inviscid fluids, the normal component of the velocity should be continuous while the tangential component may be discontinuous. However for viscous fluids both the tangential and normal components of velocity must be continuous.

The normal component of the magnetic field must be continuous at an interface, but the tangential component may be discontinuous only if one or both mediums become infinitely conducting. The tangential component of the electric field is always continuous across an interface.

Dirichlet Boundary Value Problems for Laplace Equation

The interior Dirichlet problem is the following:

If f is a continuous function prescribed on the boundary S of some finite region V, determine a function $\Phi(x, y, z)$ such that $\nabla^2 \Phi = 0$ within V and $\Phi = f$ on S.

In a similar way the exterior Dirichlet problem stands for the following:

If f is a continuous function prescribed on the boundary S of a finite simply connected region V, determine a function $\Phi(x, y, z)$ which satisfies $\nabla^2 \Phi = 0$ outside V and is such that $\Phi = f$ on S.

Thus the problem of determining the temperature at all points inside a solid when the state is steady and its temperature at the surface at all points is prescribed is an interior Dirichlet problem, while the problem of determining the potential at all points outside a conductor when the potential at all points of the conductor is known is an exterior Dirichlet problem.

The existence of the solution of the interior Dirichlet problem can be established under very general conditions. Once the existence is established, its uniqueness can be proved. For the exterior Dirichlet problem however the uniqueness proof requires some regularity conditions at infinity.

Neumann Boundary Value Problems for Laplace Equation
The interior Neumann problem is the following:

If f is a continuous function which is defined uniquely at each point of the boundary S of a finite region V, determine a function $\Phi(x, y, z)$ such that $\nabla^2\Phi = 0$ within V and its normal derivative $\partial\Phi/\partial n$ coincides with f at every point of f.

Similarly the exterior Neumann problem stands for the following:

If f is a continuous function specified at each point of the smooth boundary S of a bounded simply connected region V, find a function $\Phi(x, y, z)$ satisfying $\nabla^2\Phi = 0$ outside V and $\partial\Phi/\partial n = f$ on S.

Churchill's Boundary Value Problems for Laplace Equation
For the interior Churchill problem $\nabla^2\Phi = 0$ is satisfied at all points inside V and

$$\frac{\partial\Phi}{\partial n} + (k + 1)\Phi = 0 \tag{154}$$

at all points on S. Similarly for the exterior Churchill problem $\nabla^2\Phi = 0$ is satisfied outside V and (154) is satisfied for all points on S.

EXERCISE 6.8

1. State five problems of Dirichlet type.

2. State five problems of Neumann type.

3. State five problems of Churchill type.

4. Prove that the solutions of a certain Neumann problem can differ from one another by a constant only.

5. Prove that a necessary condition for the existence of a solution of an interior Neumann problem is that the integral of f over the boundary of S should vanish.

6. Assuming the existence of a solution of interior Dirichlet problem, prove its uniqueness.

7. Show that in the two-dimensional case, it is possible to reduce the Neumann problem to the Dirichlet problem.

7

Mathematical Modelling Through Graphs

7.1 SITUATIONS THAT CAN BE MODELLED THROUGH GRAPHS

7.1.1 Qualitative Relations in Applied Mathematics

It has been stated that "Applied Mathematics is nothing but solution of differential equations". This statement is wrong on many counts (i) Applied Mathematics also deals with solutions of difference, differential-difference, integral, integro-differential, functional and algebraic equations (ii) Applied Mathematics is equally concerned with inequations of all types (iii) Applied Mathematics is also concerned with mathematical modelling; in fact mathematical modelling has to precede solution of equations (iv) Applied Mathematics also deals with situations which cannot be modelled in terms of equations or inequations; one such set of situations is concerned with qualitative relations.

Mathematics deals with both quantitative and qualitative relationships. Typical qualitative relations are: y likes x, y hates x, y is superior to x, y is subordinate to x, y belongs to same political party as x, set y has a non-null intersection with set x; point y is joined to point x by a road, state y can be tansformed into state x, team y has defeated team x, y is father of x, course y is a prerequisite for course x, operation y has to be done before operation x, species y eats species x, y and x are connected by an airline, y has a healthy influence on x, any increase of y leads to a decrease in x, y belongs to same caste as x, y and x have different nationalities and so on.

Such relationships are very conveniently represented by graphs where a graph consists of a set of vertices and edges joining some or all pairs of these vertices. To motivate the typical problem situations which can be modelled through graphs, we consider the first problem so historically modelled viz. the problem of seven bridges of Königsberg.

7.1.2 The Seven Bridges Problem

There are four land masses A, B, C, D which are connected by seven bridges numbered 1 to 7 across a river (Figure 7.1). The problem is to start from any point in one of the land masses, cover each of the seven bridges once and once only and return to the starting point.

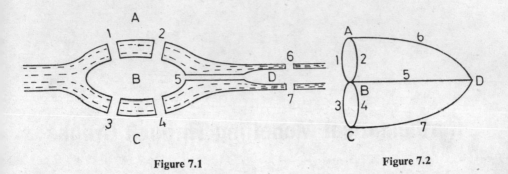

Figure 7.1   Figure 7.2

There are two ways of attacking this problem. One method is to try to solve the problem by walking over the bridges. Hundreds of people tried to do so in their evening walks and failed to find a path satisfying the conditions of the problem. A second method is to draw a scale map of the bridges on paper and try to find a path by using a pencil.

It is at this stage that concepts of mathematical modelling are useful. It is obvious that the sizes of the land masses are unimportant, the lengths of the bridges or even whether these are straight or curved are irrelevant. What is relevant information is that A and B are connected by two bridges 1 and 2, B and C are connected by two bridges 3 and 4, B and D are connected by one bridge number 5, A and D are connected by bridge number 6 and C and D are connected by bridge number 7. All these facts are represented by the graph with four vertices and seven edges in Figure 7.2. If we can trace this graph in such a way that we start with any vertex and return to the same vertex and trace every edge once and once only without lifting the pencil from the paper, the problem can be solved. Again trial and error method cannot be satisfactorily used to show that no solution is possible.

The number of edges meeting at a vertex is called the degree of that vertex. We note that the degrees of A, B, C, D-are 3, 5, 3, 3 respectively and each of these is an odd number. If we have to start from a vertex and return to it, we need an even number of edges at that vertex. Thus it is easily seen that Königsberg bridges problem cannot be solved.

This example also illustrates the power of mathematical modelling. We have not only disposed of the seven-bridges problem, but we have discovered a technique for solving many problems of the same type.

7.1.3 Some Types of Graphs
A graph is called *complete* if every pair of its vertices is joined by an edge (Figure 7.3(a)).

A graph is called a *directed graph* or a *digraph* if every edge is directed with an arrow. The edge joining A and B may be directed from A to B or from B to A. If an edge is left undirected in a digraph, it will be assumed to be directed both ways (Figure 7.3(b)).

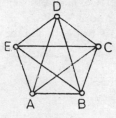

Figure 7.3a

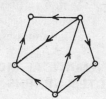

Figure 7.3b

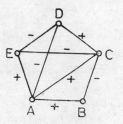

Figure 7.3c

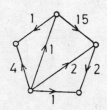

Figure 7.3d

A graph is called a *signed graph* if every edge has either a plus or minus sign associated with it (Figure 7.3(c)).

A digraph is called a *weighted digraph* if every directed edge has a weight (giving the importance of the edge) associated with it (Figure 7.3(d)). We may also have digraphs with positive and negative numbers associated with edges. These will be called *weighted signed digraphs*.

7.1.4 Nature of Models in Terms of Graphs

In all the applications we shall consider, the length of the edge joining two vertices will not be relevant. It will not also be relevant whether the edge is straight or curved. The relevant facts would be (a) which edges are joined; (b) which edges are directed and in which direction(s); (c) which edges have positive or negative signs associated with them; (d) which edges have weights associated with them and what these weights are.

EXERCISE 7.1

1. In the Königsberg problem suggest deletion or addition of minimum number of bridges which may lead to a solution of the problem.

2. Show that in any graph, the sum of local degrees of all the vertices is an even number. Deduce that a graph has an even number of odd vertices.

3. Three houses A, B, C have to be connected with three utilities a, b, c by separate wires lying in the same plane and not crossing one another. Explain why this is not possible.

4. Each of the four neighbours has connected his house with the other three houses by paths which do not cross. A fifth man builds a house nearby. Prove that (a) he cannot connect his house with all others by non-intersecting paths (b) he can however connect with three of the houses.

5. A graph is called regular if each of its vertices has same degree r. Draw regular graphs with 6 vertices and degree 5, 4 and 3.

6. Show that in Königsberg, four one-way bridges will be enough to connect the four land masses.

7.2 MATHEMATICAL MODELS IN TERMS OF DIRECTED GRAPHS

7.2.1 Representing Results of Tournaments

The graph (Figure 7.4) shows that

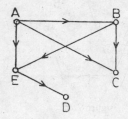

Figure 7.4

(i) Team A has defeated teams B, C, E.

(ii) Team B has defeated teams C, E.

(iii) Team E has defeated D.

(iv) Matches between A and D, B and D, C and D and C and E have yet to be played.

7.2.2 One-Way Traffic Problems

The road map of a city can be represented by a directed graph. If only one-way traffic is allowed from point a to point b, we draw an edge directed from a to b. If traffic is allowed both ways, we can either draw two edges, one directed from a to b and the other directed from b to a or simply draw an undirected edge between a and b. The problem is to find whether we can introduce one-way traffic on some or all of the roads without preventing persons from going from any point of the city to any other point. In other words, we have to find when the edges of a graph can be given direction in such a way that there is a directed path from any vertex to every other. It is easily seen that one-way traffic on the road DE cannot be introduced without disconnecting the vertices of the graph (Figure 7.5).

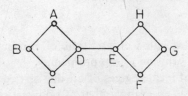

Figure 7.5(a)

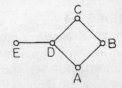

Figure 7.5(b)

In Figure 7.5(a), DE can be regarded as a bridge connecting two regions of the town. In Figure 7.5(b) DE can be regarded as a blind street on which a two-way traffic is necessary. Edges like DE are called *separating edges*, while other edges are called *circuit edges*. It is necessary that on separating edges, two-way traffic should be permitted. It can also be shown that this is sufficient. In other words, the following theorem can be established:

If G is an undirected connected graph, then one can always direct the circuit edges of G and leave the separating edges undirected (or both way directed) so that there is a directed path from any given vertex to any other vertex.

7.2.3 Genetic Graphs

In a genetic graph, we draw a directed edge from A to B to indicate that B is the child of A. In general each vertex will have two incoming edges, one from the vertex representing the father and the other from the vertex representing the mother. If the father or mother is unknown, there may be less then two incoming edges. Thus in a genetic graph, the local degree of incoming edges at each vertex must be less than or equal to two. This is a necessary condition for a directed graph to be a genetic graph, but it is not a sufficient condition. Thus Figure 7.6 does not give a genetic graph inspite of the fact that the number of incoming edges at each vertex does not exceed two. Suppose A_1 is male, then A_2 must be female, since A_1, A_2 have a child B_1. Then A_3 must be male, since A_2, A_3 have a child B_2. Now A_1, A_3 being both males cannot have a child B_3.

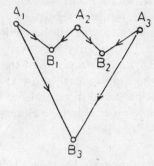

Figure 7.6

7.2.4 Senior-Subordinate Relationship

If a is senior to b, we write aSb and draw a directed edge from a to b. Thus the organisational structure of a group may be represented by a graph like the following [Figure 7.7].

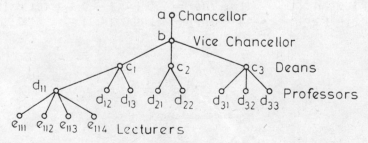

Figure 7.7

The relationship S satisfies the following properties:

(i) $\sim(aSa)$ i.e. no one is his own senior

(ii) $aSb = \sim (bSa)$ i.e. a is senior to b implies that b is not senior to a

(iii) $aSb, bSc \Rightarrow aSc$ i.e. if a is senior to b and b is senior to c, then a is senior to c.

The following theorem can easily be proved: "The necessary and sufficient condition that the above three requirements hold is that the graph of an organisation should be free of cycles"

We want now to develop a *measure for the status* of each person. The status $m(x)$ of the individual should satisfy the following reasonable requirements.

(i) $m(x)$ is always a whole number

(ii) If x has no subordinate, $m(x) = 0$

(iii) If, without otherwise changing the structure, we add a new individual subordinate to x, then $m(x)$ increases

(iv) If, without otherwise changing the structure, we move a subordinate of a to a lower level relative to x, then $m(x)$ increases.

A measure satisfying all these criteria was proposed by Harary. We define the level of seniority of x over y as the length of the shortest path from x to y. To find the measure of status of x, we find n_1, the number of individuals who are one level below x, n_2 the number of individuals who are two levels below x and in general, we find n_k the number of individuals who are k levels below x. Then the Harary measure $h(x)$ is defined by

$$h(x) = \sum_k k n_k \qquad (1)$$

It can be shown that among all the measure which satisfy the four requirements given above, Harary measure is the least.

If however, we define the level of senority of x over y as the length of the longest path from x to y, and then find $H(x) = \sum_k k n_k$, we get another measure which will be the largest among all measures satisfying the four requirements. For Figure 7.8, we get

$$h(a) = 1.2 + 4.2 + 2.3 = 16 \qquad H(a) = 1.1 + 3.2 + 2.3 + 2.4 = 21$$
$$h(b) = 1.3 + 2.4 \qquad = 11 \qquad H(b) = 2.1 + 2.2 + 2.3 + 1.4 = 16$$
$$h(c) = 1.2 + 1.2 \qquad = 4 \qquad H(c) = 1.1 + 1.2 + 1.3 \qquad = 6$$

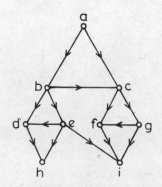

Figure 7.8

$h(d) = 1.1$		$= 1$	$H(d) = 1.1$		$= 1$
$h(e) = 1.3$		$= 3$	$H(e) = 1.2 + 2.1$		$= 4$
$h(f) = 1.1$		$= 1$	$H(f) = 1.1$		$= 1$
$h(g) = 1.2$		$= 2$	$H(g) = 1.2$		$= 2$
$h(k)$		$= 0$	$H(k)$		$= 0$
$h(l)$		$= 0$	$H(l)$		$= 0$

7.2.5 Food Webs

Here aSb if a eats b and we draw a directed edge from a to b. Here also $\sim$ (aSa) and $aSb \Rightarrow \sim(bSa)$. However the transitive law need not hold. Thus consider the food web in Fig. 7.9. Here fox eats bird, bird eats grass, but fox does not eat grass.

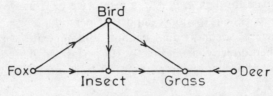

Figure 7.9

We can however calculate measure of the status of each species in this food web by using (1) $h(\text{bird}) = 2$, $h(\text{fox}) = 4$, $h(\text{insect}) = 1$, $h(\text{grass}) = 0$, $h(\text{deer}) = 1$.

7.2.6 Communication Networks

A directed graph can serve as a model for a communication network. Thus consider the network given in Figure 7.10. If an edge is directed from a to b, it means that a can communicate with b. In the given network e can communicate directly with b, but b can communicate with e only indirectly through c and d. However every individual can communicate with every other individual.

Our problem is to determine the importance of each individual in this network. The importance can be measured by the fraction of the messages on an average that pass through him. In the absence of any other knowledge, we can assume that

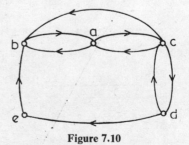

Figure 7.10

if an individual can send message direct to n individuals, he will send a message to any one of them with probability $1/n$. In the present example, the communication probability matrix is

$$
\begin{array}{ccccc}
 & a & b & c & d & e \\
\begin{array}{c} a \\ b \\ c \\ d \\ e \end{array} &
\left[\begin{array}{ccccc}
0 & 1/2 & 1/2 & 0 & 0 \\
1/2 & 0 & 1/2 & 0 & 0 \\
1/3 & 1/3 & 0 & 1/3 & 0 \\
0 & 0 & 1/2 & 0 & 1/2 \\
0 & 1 & 0 & 0 & 0
\end{array}\right]
\end{array}
\tag{2}
$$

No individual is to send a message to himself and so all diagonal elements are zero. Since all elements of the matrix are non-negative and the sum of elements of every row is unity, the matrix is a stochastic matrix and one of its eigenvalues is unity. The corresponding normalised eigenvector is $[11/45, 13/45, 3/10, 1/10, 1/15]$. In the long run, these fractions of messages will pass through a, b, c, d, e respectively. Thus we can conclude that in this network, c is the most important person.

If in a network, an individual cannot communicate with every other individual either directly or indirectly, the Markov chain is not ergodic and the process of finding the importance of each individual breaks down.

7.2.7 Matrices Associated with a Directed Graph

For a directed graph with n vertices, we define the $n \times n$ matrix $A = (a_{ij})$ by $a_{ij} = 1$ if there is an edge directed from i to j and $a_{ij} = 0$ if there is no edge directed from i to j. Thus the matrix associated with the graph of Figure 7.11 is given by

$$
A = \begin{array}{c c c c c}
 & 1 & 2 & 3 & 4 \\
\begin{array}{c} 1 \\ 2 \\ 3 \\ 4 \end{array} &
\left[\begin{array}{cccc}
0 & 1 & 1 & 0 \\
1 & 0 & 1 & 0 \\
1 & 1 & 0 & 0 \\
1 & 0 & 1 & 0
\end{array}\right]
\end{array}
\tag{3}
$$

We note that (i) the diagonal elements of the matrix are all zero (ii) the number of non-zero elements is equal to the number of edges (iii) the number of non-zero elements in any row is equal to the local outward degree of the vertex corresponding to the row (iv) the number of non-zero elements in a column is equal to the local inward degree of the vertex corresponding to the column. Now

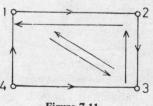

Figure 7.11

$$
A^2 = \begin{array}{c} \\ 1 \\ 2 \\ 3 \\ 4 \end{array}
\begin{array}{cccc} 1 & 2 & 3 & 4 \end{array}
\left[\begin{array}{cccc}
2 & 1 & 1 & 0 \\
1 & 2 & 1 & 0 \\
1 & 1 & 2 & 0 \\
1 & 2 & 1 & 0
\end{array} \right] = (a_{ij}^{(2)}) \tag{4}
$$

The element $a_{ij}^{(2)}$ gives the number of 2-chains from i to j. Thus from vertex 2 to vertex 1, there are two 2-chains viz. via vertex 3 and vertex 4. We can generalise this result in the form of a theorem viz. "The element $a_{ij}^{(2)}$ of A^2 gives the number of 2-chains i.e. the number of paths with two-edges from vertex i to vertex j".

The theorem can be further generalised to "The element $a_{ij}^{(m)}$ of A^m gives the number of m-chains i.e. the number of paths with m edges from vertex i to vertex j". It is also easily seen that "The ith diagonal element of A^2 gives the number of vertices with which i has symmetric relationship".

From the matrix A of a graph, a symmetric matrix S can be generated by taking the elementwise product of A with its transpose so that in our case

$$
S = A \times A^T = \left[\begin{array}{cccc}
0 & 1 & 1 & 0 \\
1 & 0 & 1 & 0 \\
1 & 1 & 0 & 0 \\
1 & 0 & 1 & 0
\end{array} \right] \times \left[\begin{array}{cccc}
0 & 1 & 1 & 1 \\
1 & 0 & 1 & 0 \\
1 & 1 & 0 & 1 \\
0 & 0 & 0 & 0
\end{array} \right] = \left[\begin{array}{cccc}
0 & 1 & 1 & 0 \\
 & 0 & 1 & 0 \\
1 & 1 & 0 & 0 \\
0 & 0 & 0 & 0
\end{array} \right]
$$
$$\tag{5}$$

S obviously is the matrix of the graph from which all unreciprocated connections have been eliminated. In the matrix S (as well as in S^2, S^3, ...) the elements in the row and column corresponding to a vertex which has no symmetric relation with any other vertex are all zero.

7.2.8 Application of Directed Graphs to Detection of Cliques

A subset of persons in a socio-psychological group will be said to form a clique if (i) every member of this subset has a symmetrical relation with every other member of this subset (ii) no other group member has a symmetric relation with all the members of the subset (otherwise it will be included in the clique) (iii) the subset has at least three members.

If other words a clique can be defined as a maximal completely connected subset of the original group, containing at least three persons. This subset should not be properly contained in any larger completely connected subset.

It the group consists of n persons, we can represent the group by n vertices of a graph. The structure is provided by persons knowing or being connected to other persons. If a person i knows j, we can draw a directed edge from i to j. If i knows j and j knows i, then we have a symmetrical relation between i and j.

With this interpretation, the graph of Figure 7.11 shows that persons 1, 2, 3 form a clique. With very small groups, we can find cliques by carefully observing the corresponding graphs. For larger groups analytical methods based on the following results are useful: (i) i is a member of a clique if the ith diagonal element of S^3 is different from zero. (ii) If there is only one clique of k members in the group, the corresponding k elements of S^3 will be $(k-1)(k-2)/2$ and the rest of the diagonal elements will be zero. (iii) If there are only two cliques with k and m members respectively and there is no element common to these cliques, then k elements of S^3 will be $(k-1)(k-2)/2$, m elements of S^3 will be $(m-1)(m-2)/2$ and the rest of the elements will be zero. (iv) If there are m disjoint cliques with k_1, k_2, ..., k_m members, then the trace of S^3 is $\frac{1}{2} \sum_{i=1}^{m} k_i(k_i-1)(k_i-2)$. (v) A member is non-cliquical if only if the corresponding row and column of $S^2 \times S$ consists entirely of zeros.

EXERCISE 7.2

1. Show that the graph of Figure 7.12 is a possible genetic graph if and only if n is even.

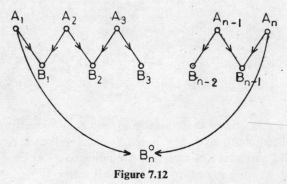

Figure 7.12

2. For each of the following communication networks, set up the corresponding transition probability matrix and find the importance of each member in the network.

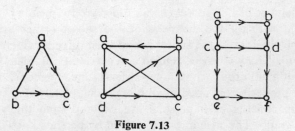

Figure 7.13

3. An intelligence officer can communicate with each of his n subordinates and each subordinate can communicate with him, but the subordinates cannot

communicate among themselves. Draw the graph and find the importance of each subordinate relative to the officer.

4. Find the Harary measure for each individual in the organisational graphs of Figure 7.14.

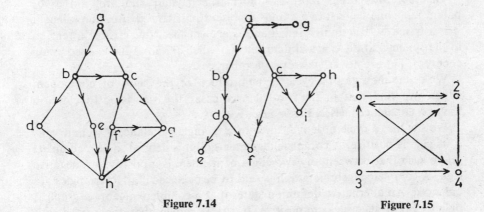

Figure 7.14   **Figure 7.15**

5. In Exercise 4, find the measure if the definition of level is based (i) in the longest number of steps between two persons (ii) on the average of the shortest and longest number of steps between two persons.

6. Find the eigenvector corresponding to the unit eigenvalue of matrix (2).

7. Prove all the theorems stated in Section 7.2.7.

8. Prove all the theorems stated in Section 7.2.8.

9. Write the matrix A associated with the graph of Figure 7.15. Find A^2, A^3, A^4, S, S^2, S^3, and verify the theorems of Sections 7.2.7 and 7.2.8.

10. Enumerate all possible four-cliques.

7.3 MATHEMATICAL MODELS IN TERMS OF SIGNED GRAPHS

7.3.1 Balance of Signed Graphs

A signed (or an algebraic) graph is one in which every edge has a positive or negative sign associated with it. Thus the four graphs of Figure 7.16 are signed graphs. Let positive sign denote friendship and negative sign denote enemity, then in graph (i) A is a friend of both B and C and B and C are

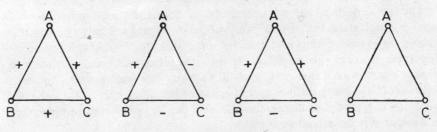

Figure 7.16

also friends. In graph (ii) A is friend of B and A and B are both jointly enemies of C. In graph (iii), A is a friend of both B and C, but B and C are enemies. In graph (iv) A is an enemy of both B and C, but B and C are not friends.

The first two graphs represents normal behaviour and are said to be balanced, while the last two graphs represent unbalanced situations since if A is a friend both B and C and B and C are enemies, this creates a tension in the system and there is a similar tension when B and C have a common enemy A, but are not friends of each other.

We define the sign of a cycle as the product of the signs of component edges. We find that in the two balanced cases, this sign is positive and in the two unbalanced cases, this is negative.

We say that a cycle of length three or a triangle is balanced if and only if its sign is positive. A complete algebraic graph is defined to be a complete graph such that between any two edges of it, there is a positive or negative sign. A complete algebraic graph is said to be balanced if all its triangles are balanced. An alternative definition states that a complete algebraic graph is balanced if all its cycles are positive. It can be shown that the two definitions are equivalent.

A graph is locally balanced at a point a if all the cycles passing through a are balanced. If a graph is locally balanced at all points of the graph, it will obviously be balanced. A graph is defined to be m-balanced if all its cycles of length m are positive. For an incomplete graph, it is preferable to define it to be balanced if all its cycles are positive. The definition in terms of triangle is not satisfactory, as there may be no triangles in the graph.

7.3.2 Structure Theorem and Its Implications

Theorem. The following four conditions are equivalent:

(i) The graph is balanced i.e. every cycle in it is positive.

(ii) All closed line-sequences in the graph are positive i.e. any sequence of edges starting from a given vertex and ending on it and possibly passing through the same vertex more than once is positive.

(iii) Any two line-sequences between two vertices have the same sign.

(iv) The set of all points of the graph can be partitioned into two disjoint sets such that every positive sign connects two points in the same set and every negative sign connects two points of different sets.

The last condition has an interesting interpretation with possibility of application. It states that if in a group of persons there are only two possible relationships viz. liking and disliking and if the algebraic graph representing these relationships is balanced, then the group will break up into two separate parties such that persons within a party like one another, but each person of one party dislikes every person of the other party. If a balanced situation is regarded as stable, this theorem can be interpreted to imply that a two-party political system is stable.

7.3.3 Antibalance and Duobalance of a Graph

An algebraic graph is said to be antibalanced if every cycles in it has an even number of positive edges. The concept can be obtained from that of a balanced graph by changing the signs of the edges. It will then be seen that an algebraic graph is antibalanced if and only if its vertices can be separated into two disjoint classes, such that each negative edge joins two vertices of the same class and each positive edge joins persons from different classes.

A signed graph is said to be duobalanced if it is both balanced and antibalanced.

7.3.4 The Degree of Unbalance of a Graph

For many purposes it is not enough to know that a situation is unbalanced. We may he interested in the degree of unbalance and the possibility of a balancing process which may enable one to pass from an unbalanced to a balanced graph. The possibility is interesting as it can give an approach to group dynamics and demonstrate that methods of graph theory can be applied to dynamic situations also.

Cartwright and Harary define the degree of balance of a group G to be the ratio of the positive cycles of G to the total number of cycles in G. This balance index obviously lies between 0 and 1. G_1 has six negative triangles viz (abc), (ade), (bcd), (bce), (bde), (cde) and has four positive triangles. G_2 has four negative triangles viz (abc), (abd), (bce) and (bde) and six positive triangle. The degree of balance of G_1 is therefore less than the degree of balance of G_2.

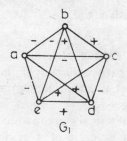 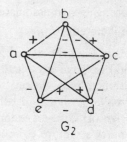 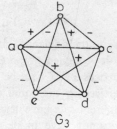

G_1 G_2 G_3

Figure 7.17

However in order to get a balanced graph from G_1, we have to change the sign of only two edges viz. bc and de and similarly to make G_2 balanced we have to change the signs of two edges viz bc and bd. From this point of view both G_1 and G_2 are equally unbalanced.

Abelson and Rosenberg therefore gave an alternative definition. They defined the degree of unbalance of an algebraic graph as the number of the smallest set of edges of G whose change of sign produces a balanced graph.

The degree of an antibalanced complete algebraic graph (i.e. of a graph all of whose triangles are negative) is given by $[n(n - 2) + k]/4$ where $k = 1$ if n is odd and $k = 0$ if n is even. It has been conjectured that the degree of unbalancing of every other complete algebraic graph is less than or equal to this value.

EXERCISE 7.3

1. State which of the following graphs are balanced. If balanced, find the decomposition guaranteed by the structure theorem. If unbalanced, find the degree of unbalance.

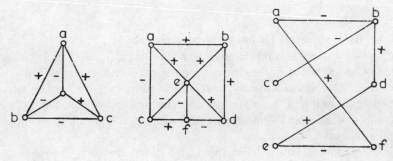

Figure 7.18

2. Draw some antibalanced graphs and verify the structure theorems for them.

3. The adjacency matrix of a signed graph is defined as follows:

$a_{ij} = 1$ if there is $+$ sign associated with edge i, j

$\quad\ = -1$ if there is $-$ sign associated with edge i, j

$\quad\ = 0$ if there is no edge i, j.

Write the adjacency matrices of the four signed graphs is Figure 7.18.

4. A signed graph G is said to have an idealised party structure if the vertices of G can be partitioned into classes so that all edges joining the vertices in the same class have $+$ sign and all edges joining vertices in different sets have negative sign (a) Give an example of a signed graph which does not have an idealised party structure (b) Give an example of a graph which is not balanced but which has an idealised party structure.

5. Show that a signed graph has an idealised party structure if and only no circuit has exactly one $-$ sign.

6. Show that if all cycles of a signed graph are positive, then all its cycles are also positive. State and prove its converse also.

7.4 MATHEMATICAL MODELLING IN TERMS OF WEIGHTED DIGRAPHS

7.4.1 Communication Networks with Known Probabilities of Communication

In the communication graph of Figure 7.10, we know that a can communi-

cate with both *b* and *c* only and
in the absence of any other knowledge, we assigned equal probabilities
to *a*'s communicating with *b* or *c*.
However we may have a priori knowledge that *a*'s chances of communicating with *b* and *c* are in the ratio
$3 : 2$, then we assign probability .6 to
a's communicating with *b* and .4 to
a's communicating with *c*. Similarly
we can associate a probability with
every directed edge and we get the

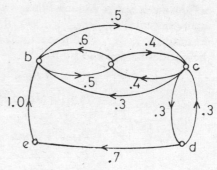

Figure 7.19

weighted digraph (Figure 7.19) with the associated matrix

$$
B = \begin{array}{c} \\ a \\ b \\ c \\ d \\ e \end{array}
\begin{array}{ccccc}
a & b & c & d & e \\
\left[\begin{array}{ccccc}
0 & 0.6 & 0.4 & 0 & 0 \\
0.5 & 0 & 0.5 & 0 & 0 \\
0.4 & 0.3 & 0 & 0.3 & 0 \\
0 & 0 & .3 & 0 & 0.7 \\
0 & 1.0 & 0 & 0 & 0
\end{array}\right]
\end{array}
\qquad (6)
$$

We note that the elements are all non-negative and the sum of the elements
of every row is unity so that B is a stochastic matrix and unity is one of its
eigenvalues. The eigenvector corresponding to this eigenvalues will be
different from the eigenvector found in Section 7.2.6 and so the relative
importance of the individuals depends both on the directed edges as well as
on the weights associated with the edges.

7.4.2 Weighted Digraphs and Markov Chains

A Markovian system is characterised by a transition probability matrix. Thus
if the states of a system are represented by 1, 2, . . ., n and p_{ij} gives the
probability of transition from the ith state to jth state, the system is characterised by the transition probability matrix (t.p.m)

$$
T = \begin{bmatrix}
p_{11} & p_{12} & .. & p_{1j} & ... & p_{1n} \\
p_{21} & p_{22} & .. & p_{2j} & ... & p_{2n} \\
\cdots & \cdots & \cdots & \cdots & \cdots & \cdots \\
p_{i1} & p_{i2} & .. & p_{ij} & .. & p_{in} \\
\cdots & \cdots & \cdots & \cdots & \cdots & \cdots \\
p_{n1} & p_{n2} & .. & p_{nj} & .. & p_{nn}
\end{bmatrix}
\qquad (7)
$$

Since $\sum_{i=1}^{n} p_{ij}$ represents the probability of the system going from ith state to

any other state or of remaining in the same state, this sum must be equal to
unity. Thus the sum of elements of every row of a t.p.m. is unity.

Consider a set of N such Markov systems where N is large and suppose at any instant NP_1, NP_2, . . ., NP_n of these $(P_1 + P_2 + \ldots + P_n = 1)$ are in states 1, 2, 3, . . ., n respectively. After one step, let the proportions in these states be denoted by P_1', P_2', . . ., P_n', then

$$P_1' = P_1p_{11} + P_2p_{21} + P_3p_{31} + \ldots \ldots + P_np_{n1}$$
$$P_2' = P_2p_{12} + P_2p_{22} + P_3p_{32} + \ldots \ldots + P_np_{n2}$$
$$\ldots \ldots \ldots \ldots \ldots \ldots \ldots \ldots \ldots \ldots \ldots \ldots \ldots \tag{8}$$
$$P_n' = P_1p_{1n} + P_2p_{2n} + P_3p_{3n} + \ldots \ldots + P_np_{nn}$$

or
$$P' = PT \tag{9}$$

where P and P' are row matrices representing the proportions of systems in various states before and after the step and T is the t.p.m.

We assume that the system has been in operation for a long time and the proportions P_1, P_2, . . ., P_n have reached equilibrium values. In this case

$$P = PT \quad \text{or} \quad P(I - T) = 0, \tag{10}$$

where I is the unit matrix. This represents a system of n equations for determining the equilibrium values of P_1, P_2, . . ., P_n. If the equations are consistent, the determinant of the coefficient must vanish i.e. $|T - I| = 0$. This requires that unity must be an eigenvalue of T. However this, as we have seen already is true. This shows that an equilibrium state is always possible for a Markov chain.

A Markovian system can be represented by a weighted directed graph. Thus consider the Markovian system with the stochastic matrix

$$
\begin{array}{c c c c c}
 & a & b & c & d \\
a & \begin{bmatrix} 0.2 & 0.8 & 0 & 0 \\ 0.3 & 0.6 & 0.1 & 0 \\ 0.2 & 0.4 & 0.3 & 0.1 \\ 0 & 0 & 0 & 1 \end{bmatrix}
\end{array}
$$

Its weighted digraph is given in Figure 7.20.

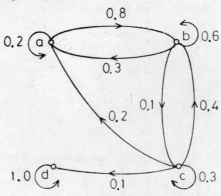

Figure 7.20

In this example d is an absorbing state or a state of equilibrium. Once a system reaches the state d, it stays there for ever.

It is clear from Figure 7.20, that in whichever state, the system may start, it will ultimately end in state d. However the number of steps that may be required to reach d depends on chance. Thus starting from c, the number of steps to reach d may be 1, 2, 3, 4, . . .; starting from b the number of steps to reach d may be 2, 3, 4, . . . and starting for a, the number of steps may be 3, 4, 5, . . . In each case, we can find the probability that the number of steps required is n and then we can find the expected number of steps to reach it.

Thus for the matrix

$$
\begin{array}{cc}
 & a \qquad\quad b \\
\begin{array}{c} a \\ b \end{array} &
\left[\begin{array}{cc} 1 & 0 \\ 1/3 & 2/3 \end{array} \right]
\end{array}
\tag{12}
$$

a is an absorbing state. Starting from b, we can reach a in 1, 2, 3, . . ., n steps with probabilities $(1/3)$, $(1/3)(2/3)$, $(1/3)(2/3)^2$, . . ., $(1/3)(2/3)^{n-1}$, . . ., so that the expected number of steps is

$$
\sum_{n=1}^{\infty} n\frac{1}{3}\left(\frac{2}{3}\right)^{n-1} = 3
\tag{13}
$$

7.4.3 General Communication Networks

So for we have considered communication networks in which the weight associated with a directed edge represents the probability of communication along that edge. We can however have more general networks e.g.

(a) for communication of messages where the directed edge represents the channel and the weight represents the capacity of the channel say in bits per second

(b) for communication of gas in pipelines where the weights are the capacities, say in gallons per hour

(c) communication roads where the weights are the capacities is cars per hour.

An interesting problem is to find the maximum flow rate, of whatever is being communicated, from any vertex of the communication network to any other. Useful graph-theoretic algorithms for this have been developed by Elias. Feinstein and Shannon as well as by Ford and Fulkerson.

7.4.4 More General Weighted Digraphs

In the most general case, the weight associated with a directed edge can be positive or negative. Thus Figure 7.21 means that a unit change at vertex 1 at time t causes changes of -2 units at vertex 2, of 2 units at vertex 4 and of 3 units at vertex 5 at time $t + 1$. Similarly a change of 1 unit

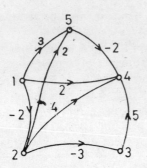

Figure 7.21

at vertex 2 causes a change of -3 units at 3 vertex, 4 units at vertex 4 and of 2 units at vertex 5 and so on. Given the values at all vertices at time t, we can find the values at times $t + 1, t + 2, t + 3, \ldots$ The process of doing this systematically is known as the pulse rule.

These general weighted digraphs are useful for representing energy flows, monetary flows and changes in environmental conditions.

7.4.5. Signal Flow Graphs

The system of algebraic equations

$$\begin{aligned} x_1 &= 4y_0 + 6x_2 - 2x_3 \\ x_2 &= 2y_0 - 2x_1 + 2x_3 \\ x_3 &= 2x_1 - 2x_2 \end{aligned} \tag{14}$$

can be represented by the weighted digraph in Figure 7.22. For solving for x_1, we successively eliminate x_3 and x_2 to get the graphs in Figure 7.23 and finally we get

$$x_1 = 4y_0$$

We can similarly represent the solution of any number of linear equations graphically.

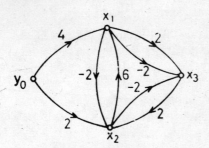

Figure 7.22

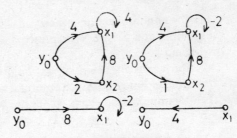

Figure 7.23

7.4.5 Weighted Bipartitic Digraphs and Difference Equations

Consider the system of difference equations

$$\begin{aligned} x_{t+1} &= a_{11}x_t + a_{12}y_t + a_{13}z_t \\ y_{t+1} &= a_{21}x_t + a_{22}y_t + a_{23}z_t \\ z_{t+1} &= a_{31}x_t + a_{32}y_t + a_{33}z_t \end{aligned} \tag{15}$$

This can be represented by a weighted bipartitic digraph (Figure 7.24). The weights can be positive or negative.

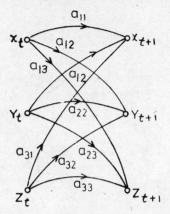

Figure 7.24

EXERCISE 17.4

1. A machine can be in any one of the states a, b, c. The transitions between states are governed by the transition probability matrix

$$
\begin{array}{c c c c}
 & a & b & c \\
a & \begin{bmatrix} 1 & 0 & 0 \\ 1/2 & 0 & 1/2 \\ 1/3 & 1/3 & 1/3 \end{bmatrix}
\end{array}
\qquad (16)
$$

Draw the weighted digraph and find the limiting probabilities for the machine to be found in each of the three states.

2. The entropy of a Markov machine is defined by

$$
H = \sum_{i=1}^{n} P_i H_i = - \sum_{i=1}^{n} \sum_{j=1}^{n} P_i p_{ij} \ln p_{ij} \qquad (17)
$$

Show that

(a) When

$$
T = 2 \quad
\begin{array}{c c c c}
 & 1 & 2 & 3 \\
1 & \begin{bmatrix} 1/4 & 3/4 & 0 \\ 3/4 & 0 & 1/4 \\ 1/8 & 3/4 & 1/8 \end{bmatrix}
\end{array}
$$

$$
P_1 = 0.449, \quad P_2 = 0.429, \quad P_3 = 0.122
$$
$$
H_1 = 0.811, \quad H_2 = 0.811, \quad H_3 = 1.663
$$

(b) When

$$
T = \quad
\begin{array}{c c c c c}
 & 1 & 2 & 3 & 4 \\
1 & \begin{bmatrix} 0 & 0.6 & 0.4 & 0 \\ 0 & 0.6 & 0.4 & 0 \\ 0.3 & 0 & 0 & 0.7 \\ 0.3 & 0 & 0 & 0.7 \end{bmatrix}
\end{array}
$$

$$P_1 = 6/35, \quad P_2 = 9/35,$$
$$P_3 = 6/35, \quad P_4 = 14/35$$
$$H = 0.92$$

3. In a panel survey, a person is asked a question to which he can answer 'Yes' or 'No'. In the next survey, the probability of his being in state 1 (Yes) or state 2 (No) is given by

$$
\begin{array}{cc}
 & 1 \qquad\quad 2 \\
\begin{array}{c} 1 \\ 2 \end{array} &
\left[\begin{array}{cc} 1 - \alpha & \alpha \\ \beta & 1 - \beta \end{array} \right]
\end{array}
\tag{18}
$$

Show that

(a) $p_1(t + 1) = p_1(t)(1 - \alpha) + p_2(t)\beta$

$$p_2(t + 1) = p_1(t)\alpha + p_2(t)(1 - \beta) \tag{19}$$

(b) $p_1(t) = \dfrac{\beta}{\alpha + \beta} + (1 - \alpha - \beta)^t \left[p_1(0) - \dfrac{\beta}{\alpha + \beta} \right]$

$$p_2(t) = \dfrac{\alpha}{\alpha + \beta} + (1 - \alpha - \beta)^t \left[p_2(0) - \dfrac{\alpha}{\alpha + \beta} \right] \tag{20}$$

(c) $p_1(t)$, $p_2(t)$ approaches $\beta/(\alpha + \beta)$ and $\alpha/(\alpha + \beta)$ as $t \to \infty$ if $\alpha + \beta \leq 1$.

4. In Exercise 3, find the expected number of time units in which the system now in state 1(2) will change to state 2(1).

5. Interpret the models and results of Exercises 3 and 4 when states 1, 2 refer to

(a) a neuron being excited or not excited
(b) a machine being in working order or out of order
(c) a stimulus being or not being available in a learning situation
(d) a daily wage worker being employed or not employed.

6. Give the graphical solution of

$$
\begin{aligned}
x_1 - 2x_2 + 3x_3 &= 2 \\
3x_1 + x_2 - x_3 &= 3 \\
x_1 + 2x_2 + x_3 &= 4
\end{aligned}
\tag{21}
$$

7.5 MATHEMATICAL MODELLING IN TERMS OF UNORIENTED GRAPHS

7.5.1 Electrical Networks and Kirchoffs' Laws

For more than a hundred years after Euler solved the Königsberg problem in 1736, graph theory continued to deal with interesting puzzles only. It was in 1849 that Kirchoffs' formulation of his laws of electrical currents in graph-theoretic terms led to interest in serious applications of graph theory.

An electrical circuit (Figures 7.25a, b) consists of resistors R_1, R_2, ..., inductances L_1, L_2, ..., capacitors C_1, C_2 and batteries B_1 B_2, etc.

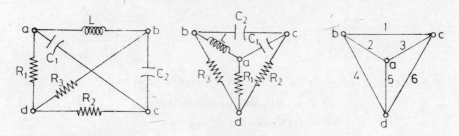

Figure 7.25

The network diagram represents two independent aspects of an electrical network. The first gives the interconnection between components and the second gives voltage-current relationship of each component. The first aspect is called network topology and can be modelled graphically. This aspect is independent of voltages and currents. The second aspects involves voltages and current and is modelled through differential equations.

For topological purposes, lengths and shapes of connections are not important and graphs of Figures 7.25(a), 7.25(b) and 7.25(c) are isomorphic.

For stating Kirchoff's laws, we need two incidence matrices accociated with the graph. If v and e denote the number of vertices and edges respectively, we define the *vertex* or *incidence* matrix $A = [a_{ij}]$ as follows:

$a_{ij} = 1$ if the edge j is incident at vertex i

$a_{ij} = 0$ if the edge j is not incident at vertex i

This consists of v rows and e columns. For graph 7.25, A is given by

$$A = \begin{array}{c} \\ a \\ b \\ c \\ d \end{array} \begin{array}{cccccc} 1 & 2 & 3 & 4 & 5 & 6 \\ \left[\begin{array}{cccccc} 0 & 1 & 1 & 0 & 1 & 0 \\ 1 & 1 & 0 & 1 & 0 & 0 \\ 1 & 0 & 1 & 0 & 0 & 1 \\ 0 & 0 & 0 & 1 & 1 & 1 \end{array}\right] \end{array} \qquad (22)$$

We note that every column has two non-zero elements.

Similarly we define the circuit matrix $B = [b_{kj}]$ as follows

$b_{kj} = 1$ if element j is in circuit k

$\quad\ = 0$ if element j is not in circuit k

The matrix B contains as many rows as there are circuits and it has e columns. In our case

$$B = \begin{array}{c} \\ 1 \\ 2 \\ 3 \\ 4 \end{array} \begin{array}{cccccc} 1 & 2 & 3 & 4 & 5 & 6 \\ \left[\begin{array}{cccccc} 1 & 1 & 1 & 0 & 0 & 0 \\ 0 & 1 & 0 & 1 & 1 & 0 \\ 0 & 0 & 1 & 0 & 1 & 1 \\ 1 & 0 & 0 & 1 & 0 & 1 \end{array}\right] \end{array} \qquad (23)$$

Now Kirchoff's laws can be written in the matrix form as follows:

$$AI = 0 \text{ (Kirchoff's current law)} \qquad (24)$$
$$BV = 0 \text{ (Kirchoff's voltage law)} \qquad (25)$$

where I is an exl column matrix giving the e currents and V is exl column matrix giving e voltages.

Matrices A and B depend on the graph only, matrices I and V depend on currents and voltages only. A and B can be written independently of I and V. Now an important question is as to how many of the components of the current and voltage vectors are independent.

It can be proved that the rank of A is $v - 1$ and the rank of B is $e - v + 1$. Thus $v - 1$ and $e - v + 1$ are the numbers of linearly independent Kirchoffs current and voltage equations.

The graph-theoretic methods can now be used to (i) establish the validity of the circuit and vertex equations and find their generalisations (ii) conditions under which unique solutions of these equations exist (iii) justify the duality procedures used in network theory (iv) develop short-cut methods for writing equations (v) develop techniques for network synthesis.

7.5.2 Lumped Mechanical Systems

If the linear graph represents a lumped mechanical system with the vertices representing rigid bodies, matrices A and B arise for Newtons' force and displacements equations respectively and $v - 1$ and $e - v + 1$ represent the number of linearly independent force and displacement equations.

7.5.3 Map-Colouring Problems

The four colour problem that every plane map, however complex, can be coloured with four colours in such a way that two neighbouring regions get different colours, challenged and fascinated mathematicians for over one hundred years till it was finally solved by Appall and Haken in 1976 by using over 1000 hours of computer time. The problem is essentially graph-theoretic since the sizes and shapes of regions are not important. That four colours are necessary is easily seen by considering the simple graph in Figure 7.26. It was the proof of the sufficiency that took more than hundred years. However the efforts

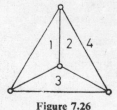

Figure 7.26

to solve this problem led to the development of many other graph-theoretic models.

Similar map-colouring problems arise for colouring of maps on surface of a sphere, a torus or other surfaces. However many of these were solved even before the simpler-looking four-colour problem was disposed of.

7.5.4 Planar Graphs

In printing of T.V. and radio circuits; we want that the wires, all lying in a plane, should not intersect. In the graph of Figure 7.27a wires appear to intersect, but we can find an isomorphic graph in Figure 7.27(b) in which edges do not intersect. A graph which is such that we can draw a graph isomorphic to it in which edges do not intersect is called a planar graph.

Figure 7.27 (a)

Figure 7.27 (b)

A complete graph with five vertices is not planar (Figure 7.28a). We can draw nine of the edges so that these do not intersect (Figure 7.28b) but however we may draw, we cannot draw all the ten edges without at least two of them intersecting. The proof of this depends on Jordan's theorem that every simple closed curve divides the plane into two regions, one inside the curve and one outside the curve. *ABCDE* in Figure 7.28(b) is a closed Jordan curve and we cannot draw three edges either inside it or outside it without intersecting.

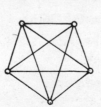

Figure 7.28 (a)

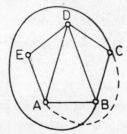

Figure 7.28 (b)

7.5.5 Euler's Formula for Polygonal Graphs

A polygonal graph with n vertices and n straight or curved edges has n vertices, n edges and two faces (one inside and one outside) so that for this graph

$$V - E + F = 2 \tag{26}$$

If we add on one edge, another polygonal region of r vertices, we increase the number of vertices by $r - 2$, the number of edges by $r - 1$ and the number of faces by 1, so that the net increases in $V - E + F$ is zero and the formula (26) remains valid. It can be shown by using the principle of induction that (26) is valid for any polygonal graph with any number of regions.

To draw the dual graph G^* of G, we take a point inside each region and draw an edge through it intersecting one of the edges of the region. It is obvious that for this dual graph the number of vertices, edges and faces is given by

$$V^* = F, \quad E = E^*, \quad F^* = V, \tag{30}$$

so that $$V^* - E^* + F^* = F - E + V = 2, \tag{31}$$

as expected.

7.5.6 Regular Solids

A polygonal graph G is said to be completely regular if both G and its dual G^* are regular i.e. if the degree of each vertex of G is the same (say ρ) and the degree of each vertex of G^* is the same (say ρ^*). From this definition, it follows

$$2E = \rho V = \rho^* F \tag{32}$$

or $$E = \frac{1}{2} \rho V, F = \frac{\rho}{\rho^*} V \tag{33}$$

Substituting (33) in (26)

$$V - \frac{1}{2}\rho V + \frac{\rho}{\rho^*}V = 2 \tag{34}$$

or $$V(2\rho + 2\rho^* - \rho\rho^*) = 4\rho^* \tag{35}$$

Since V, ρ, ρ^* are positive integers

$$2\rho + 2\rho^* - \rho\rho^* > 0 \quad \text{or} \quad (\rho - 2)(\rho^* - 2) < 4 \tag{36}$$

If $\rho > 2$, $\rho^* > 2$, the only solutions of the inequality (36) are

$$\rho = 3, \quad \rho^* = 3; \quad \rho = 3; \quad \rho^* = 4; \quad \rho = 3, \quad \rho^* = 5; \quad \rho = 4, \quad \rho^* = 3;$$
$$\rho = 5, \quad \rho^* = 3.$$

Substituting in (35) and (33), we get the table and graphs

	ρ	V	E	F	ρ^*	V^*	$E*$	F^*
(i)	3	4	6	4	3	4	6	4
(ii)	3	8	12	6	4	6	12	8
(iii)	3	20	30	12	5	12	30	20
(iv)	4	6	12	8	3	8	12	6
(v)	5	12	30	20	3	20	30	12

The corresponding graphs are given in Figure 7.29(a)-(e). It is obvious that tetrahedron graph is dual to itself, cube is dual of octahedron and Dodocahedron and Icosahedron are duals of each other.

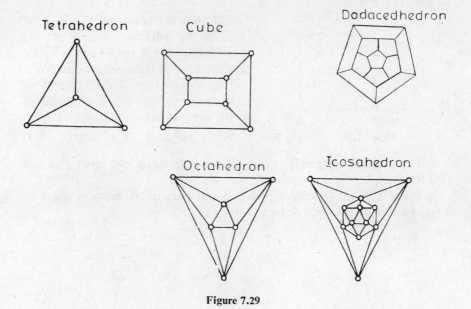

Figure 7.29

These five graphs corresponding to five **Platonic** regular solids (Figure 7.30).

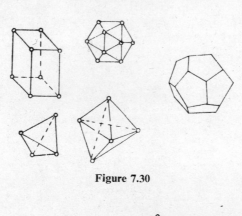

Figure 7.30

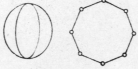

Figure 7.31

There is another solution of (36) viz. $\rho = 2$, $\rho^* = 2, 3, 4, \ldots$ The corresponding graphs G and G^* are shown in Figure 7.31.

EXERCISE 7.5

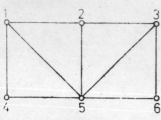

Figure 7.32

1. For the graph of Figure 7.32, write the adjacency matrix A and circuit matrix B and find their ranks. Find a set of independent circuits.

2. Prove that if the columns of matrices A and B are arranged in the same element order, then

$$AB^T = 0, \quad BA^T = 0 \qquad (37)$$

3. Draw some polygonal graphs. Draw their duals and verify (26) and (31) for them.

4. Prove that all repetitive planar graph pattern or mosaics must be formed either by triangles or by quadrangles or by hexagon.

Mathematical Modelling Through Functional, Integral, Delay-Differential and Differential-Difference Equations

8.1 MATHEMATICAL MODELLING THROUGH FUNCTIONAL EQUATIONS

8.1.1 Functional Equations

Consider the equations

$$f(x) + f(y) = f(x + y) \tag{1}$$

$$f(x)f(y) = f(x + y) \tag{2}$$

$$f(x) + f(y) = f(xy) \tag{3}$$

$$f(x)f(y) = f(xy) \tag{4}$$

Each of these involves an unknown function $f(\cdot)$, which has to be determined. Differential and integral equations also involve unknown functions, but these functions are operated by differential and integral operators. Difference equations also contain unknown functions, but here the values of these functions at equidistant points like $x - 2, x - 1, x, x + 1, x + 2$ etc. are related. In equations (1)-(4), there are no differential or integral or differencing operators, but the values of $f(\cdot)$ at two or more distinct points x, y, z, are related. Such equations are called functional equations and many important mathematical models are described in terms of these functional equations.

We can easily guess solutions of the equations $(1) - (4)$. These are

$$f(x) = Ax, \quad f(x) = e^{Bx}, \quad f(x) = C \ln x, \quad f(x) = x^D, \tag{5}$$

where A, B, C, D are arbitrary constants. The question naturally arises whether these are the most general solutions, at least among the class of continuous and differentiable functions. It can be shown that this is in fact true, but we are not proving this result here.

We can generate functional equations for many elementary functions. Thus

$$f(x) = \tan^{-1} x \Rightarrow f(x) \pm f(y) = f\left(\frac{x \pm y}{1 \mp xy}\right) \tag{6}$$

$$f(x) = \sin x, g(x) = \cos x \Rightarrow f(x \pm y) = f(x)g(y) \pm g(x)f(y) \qquad (7)$$

$$f(x) = \sin x, g(x) = \cos x \Rightarrow g(x \pm y) = g(x)g(y) \pm f(x)f(y) \qquad (8)$$

$$f(x) = \sin x \qquad\qquad \Rightarrow f^2(x) - f^2(y) = f(x + y)f(x - y). \quad (9)$$

8.1.2 Lagrange's Formula for Area of a Rectangle

The area of a rectangle of sides x and y should be a function $f(x, y)$ of x and y, which on the basis of intuitive notion of area, satisfies the functional equations

$$f(x_1 + x_2, y) = f(x_1, y) + f(x_2, y) \qquad (10)$$

$$f(x, y_1 + y_2) = f(x, y_1) + f(x, y_2). \qquad (11)$$

From (1), (5) and (10)

$$f(x, y) = A(y)x, \qquad (12)$$

where $A(y)$ is an arbitrary function of y. Again (1), (5) and (11), give

$$f(x, y) = B(x)y, \qquad (13)$$

where $B(x)$ is an arbitrary function of (x). Both (12) and (13) are satisfied by

$$f(x, y) = Cxy, \qquad (14)$$

where C is an arbitrary constant. Now we define unit area as the area of a rectangle whose length and breadth are unity. This gives $C = 1$ and we get

$$f(x, y) = xy, \qquad (15)$$

which gives the expression for the area of a rectangle with length x and breadth y. It may be noted that unlike the usual proof for the expression of area of a rectangle which is valid for rational number lengths only, the present proof applies when x and y are any real numbers.

8.1.3 Formula for Compound Interest

Let x be the principal, r be the rate of interest per unit amount per unit time and let y be the time period for which money is invested, then the final amount will be some function $f(x, y)$. The amount will be the same whether principal $x_1 + x_2$ is invested together or principals x_1, x_2, are invested separately, so that

$$f(x_1 + x_2, y) = f(x_1, y) + f(x_2, y) \qquad (16)$$

From (1), (5) and (16)

$$f(x, y) = D(y)x, \qquad (17)$$

where $D(y)$ is an arbitrary function of y. Now the final amount will be the same whether we invest amount x for period $y_1 + y_2$ or we invest principal x for period y_1 and then invest the resulting amount for period y_2 so that,

$$D(y_1 + y_2)x = D(y_2)D(y_1)x \qquad (18)$$

From (2), (5) and (18)

$$D(y) = e^{Ky} \qquad (19)$$

where K is a constant. From (17), (18) and (19)

$$f(x, y) = xe^{Ky} \tag{20}$$

Now
$$f(x, 1) = x(1 + r) \tag{21}$$

From (20) and (21)

$$f(x, y) = x(1 + r)^y, \tag{22}$$

which is the formula for the amount for principal x invested at the rate r for period y.

8.1.4 Entropy of a Probability Distribution

Let the information given by the happening of an event with probability for happening p be denoted by $f(p)$, then

$$f(p) \geqslant 0 \tag{23}$$

and
$$f(pq) = f(p) + f(q), \tag{24}$$

where p and q are the probabilities of happening of two independent events. From (3), (5), (23) and (24)

$$f(p) = c \ln p \tag{25}$$

and since we take $0 < p \leqslant 1$, $\ln p < 0$, c is negative, $c = -k$, where $k > 0$ and is arbitrary, so that

$$f(p) = -k \ln p \tag{26}$$

If there are n outcomes with probabilities $p_1, p_2, \ldots, p_n$ where

$$\sum_{i=1}^{n} p_i = 1, \qquad p_i > 0 \; \forall_i,$$

then the weighted information is given by

$$H(p_1, p_2, \ldots, p_n) = -k \sum_{i=1}^{n} p_i \ln p_i \tag{27}$$

Since $p \ln p \to 0$, as $p \to 0$, we can define $0 \ln 0 = 0$ and the expression for $H(p_1, p_2, \ldots, p_n)$ is valid even when one or more of the p_i's are zero. The function $H(p) = H(p_1, p_2, \ldots, p_n)$ is defined as the entropy of the probability distribution $P = (p_1, p_2, \ldots, p_n)$.

8.1.5 The Basic Functional Equation of Information Theory

When $n = 2$, (27) gives

$$g(p) = H(p, 1 - p) = -k[p \ln p + (1 - p) \ln (1 - p)] \tag{28}$$

This is easily seen to satisfy the functional equation

$$g(x) + (1 - x)g\left(\frac{y}{1 - x}\right) = g(y) + (1 - y)g\left(\frac{x}{1 - y}\right) \tag{29}$$

This is called the basic functional equation of information theory.

8.1.6 A Generalisation of Functional Equation of Information Theory

This is given by

$$g(x) + (1 - x)^\beta g\left(\frac{y}{1 - x}\right) = g(y) + (1 - y)^\beta g\left(\frac{x}{1 - y}\right) \qquad (30)$$

of which (29) is a special case when $\beta = 1$. It is easily verified that a general solution of (30) is

$$g(x) = h(\beta)[x^\beta + (1 - x)^\beta - 1] \qquad (31)$$

where $h(\beta)$ is any arbitrary function of β. If $h(\beta) = 1/k(\beta)$ and $k(1) = 0$, then

$$\underset{\beta \to 1}{\mathrm{Lt}}\ g(x) = \underset{\beta \to 1}{\mathrm{Lt}}\ \frac{x^\beta + (1 - x)^\beta - 1}{k(\beta)} = \frac{x \ln x + (1 - x) \ln (1 - x)}{k'(1)} \qquad (32)$$

which gives the solution (28) of section 8.1.5.

8.1.7 Another Functional Equation of Information Theory

Let

$$H_n(P) = H(p_1, p_2, \ldots, p_n) = -\sum_{i=1}^{n} p_i \ln p_i \qquad (33)$$

and

$$H_m(Q) = H(q_1, q_2, \ldots, q_m) = -\sum_{j=1}^{m} q_j \ln q_j \qquad (34)$$

be the entropies of two independent probability distributions, then the entropy of the joint probability distribution is given by

$$H_{mn}[p_1q_1, \ldots, p_nq_m] = -\sum_{j=1}^{m} \sum_{i=1}^{n} p_iq_j \ln (p_iq_i)$$

$$= -\sum_{j=1}^{m} q_j \ln q_j \sum_{i=1}^{n} p_i - \sum_{j=1}^{m} q_j \sum_{i=1}^{n} p_i \ln p_i$$

$$= H_m(Q) + H_n(P)$$

or

$$H_{mn}(PQ) = H_m(Q) + H_n(P) \qquad (35)$$

This suggests that the functional equation

$$\sum_{j=1}^{m} \sum_{i=1}^{n} f(x_iy_j) = \sum_{i=1}^{n} f(x_i) + \sum_{j=1}^{m} f(y_j), \qquad (36)$$

where

$$x_i \geqslant 0, \quad y_j \geqslant 0 \sum_{i=1}^{n} x_i = 1, \quad \sum_{j=1}^{m} y_j = 1 \qquad (37)$$

is satisfied by

$$f(x) = Ax \ln x. \qquad (38)$$

8.1.8 Functional Equations in Maximum Likelihood Estimation

Let a random sample $x_1, x_2, \ldots, x_n$ be drawn from a population with probability density function $f(x - m)$, then the likelihood function is given by

$$L(x_1, x_2, \ldots, x_n, m) = f(x_1 - m)f(x_2 - m) \ldots f(x_n - m) \qquad (39)$$

According to the principle of maximum likelihood estimation, we choose that estimate for m which maximizes L so that we get

$$\frac{f'(x_1 - m)}{f(x_1 - m)} + \frac{f'(x_2 - m)}{f(x_2 - m)} + \ldots + \frac{f'(x_n - m)}{f(x_n - m)} = 0 \tag{40}$$

or
$$g(w_1) + g(w_2) + \ldots + g(w_n) = 0, \tag{41}$$

where
$$g(w) = \frac{f'(x - m)}{f(x - m)}, \qquad w = x - m \tag{42}$$

If the sample arithmetic mean is the maximum likelihood estimator, then

$$\sum_{1=i}^{n} (x_i - m) = 0 \qquad \text{or} \qquad \sum_{i=1}^{n} w_i = 0 \tag{43}$$

One solution of functional equation (41) subject to (43) is

$$g(w) = Aw, \tag{44}$$

where A is an arbitrary constant. From (42) and (44)

$$\frac{f'(w)}{f(w)} = Aw \tag{45}$$

Integrating (45),

$$\ln f(w) = \int Aw \, dw = B + \frac{1}{2} Aw^2 \tag{46}$$

$$f(x - m) = c \exp \left[\frac{1}{2} A (x - m)^2 \right] \tag{47}$$

In particular if we take $A = -\frac{1}{\sigma^2}$, we get

$$f(x - m) = \frac{1}{\sqrt{2\pi}\sigma} \exp \left(-\frac{1}{2} \frac{(x - m)^2}{\sigma^2} \right), \tag{48}$$

if the range of x is $(-\infty, \infty)$.

8.1.9 Functional Equations Arising in Dynamic Programming

We assume that we know the optimal policy or sequence of decisions for an $(n - 1)$-stage process, then to find the optimal policy for the n-stage process, we choose the first decision arbitrarily and then use the optimal policy for the remaining $(n - 1)$ stages, starting from the result due to the first decision. The net result of combining the result of the first decision and the optimal result for the remaining $(n - 1)$ stages depends essentially on the first decision. We then choose the first decision to optimize this combined result. This procedure is according to the principle of optimal policy.

Thus let our problem be concerned with maximizing $x_1 x_2 \ldots x_n$ subject to $x_1 + x_2 + \ldots + x_n = c$ and each $x_i \geqslant 0$. The maximum value will depend on n and c and as such may be denoted by $f_n(c)$. We have to take n decisions viz. about the values of $x_1, x_2, \ldots, x_n$. Let us choose x_1 arbitrarily between o and c, then $c - x_1$ will have to be distributed into $(n - 1)$ parts so as to maximize the product of these $n - 1$ points. This maximum product is $f_{n-1}(c - x_1)$ and as such the maximum value for decision into n parts is $x_1 f_{n-1}(c - x_1)$ and this depends on x_1. We chose x_1 between 0 and c to maximize it to get the equation

$$f_n(c) = \max_{0 \leqslant x_1 \leqslant c} [(x_1 f_{n-1}(c - x_1)] \tag{49}$$

Also obviously

$$f_1(c) = c, \tag{50}$$

so that

$$f_2(c) = \max_{0 \leqslant x_1 \leqslant c} [x_1(c - x_1)] = \left(\frac{c}{2}\right)^2 \tag{51}$$

$$f_3(c) = \max_{0 \leqslant x_1 \leqslant c} \left[x_1\left(\frac{c - x_1}{2}\right)^2\right] = \left(\frac{c}{3}\right)^3 \tag{52}$$

$$f_4(c) = \max_{0 \leqslant x_1 \leqslant c} \left[x_1\left(\frac{c - x_1}{3}\right)^3\right] = \left(\frac{c}{4}\right)^4 \tag{53}$$

and so on. We can use mathematical induction to show that for all values of n

$$f_n(c) = \left(\frac{c}{n}\right)^n \tag{54}$$

Equation (50) is a new type of functional equation which expresses $f_n(c)$ in terms of knowledge of all values of $f_{n-1}(x)$ for all values of x between 0 and c. It is a functional difference equation since there is difference with respect to n and there is a relation between the value of $f_n(c)$ and a continuous set of values of $f_{n-1}(x)$.

Many multistage optimization mathematical modelling problem require the solution of such functional equations involving differencing and optimizing operators. We shall discuss these problems again in section 9.3.

EXERCISE 8.1

1. Find functional equations satisfied by

$$f(x) = \sinh x, \qquad g(x) = \cosh x, \qquad h(x) = \tanh x.$$

2. Use the method of section 8.1.2 to show that the volume of a rectangular parallelopiped is xyz. Generalise this result to n dimensional space.

3. Use the method of section 8.13 to show that if the interest is payable n times a period, then $f(x, y) = x(1 + r/n)^{ny}$. Find its limit as $n \to \infty$ and interpret the result.

4. Use the principle of optimality to show that (i) $- \sum_{i=1}^{n} p_i \ln p_i$ is maximum subject is $\sum_{i=1}^{n} p_i = c$ when $p_i = c/n$ (ii) The minimum value of $x_1 + x_2 + \ldots + x_n$ subject to $x_1 x_2 \ldots x_n = d$, $x_i > 0$ is $n d^{1/n}$.

5. Prove that (i) the volume of a rectangular parallelopiped with given perimeter is maximum when the parallelopiped is a cube (ii) the perimeter

of a rectangular parallelopiped with a given volume is minimum when the parallelopiped is a cube (iii) the arithmetic mean of n positive numbers $\geqslant$ their geometric mean $\geqslant$ their harmonic mean.

6. Verify that (i) $g(x) = -(x \ln x + (1 - x) \ln (1 - x)$ satisfies (29) (ii) The function $g(x)$ in (31) satisfies (30) (iii) the function $f(x)$ in (38) satisfies (36).

7. Find the most general continuous variate probability distribution for which (i) sample arithmetic mean (ii) the sample geometric mean (iii) the sample harmonic mean is the maximum likelihood estimator for a population parameter.

8. Find A, B, C, so that $f(x) = A(1 - x) \ln (1 - x) + B(1 - x) + C$ satisfies the functional equation

$$\sum_{j=1}^{n} \sum_{i=1}^{m} f(1 - x_i y_j) = \sum_{i=1}^{m} f(1 - x_i) + \sum_{j=1}^{n} f(1 - y_j),$$

where

$$\sum_{i=1}^{m} x_i = \sum_{j=1}^{n} y_j = 1, \quad x_i \geqslant 0, \quad y_j \geqslant 0.$$

9. Show that the Fermi-Dirac entropy function

$$f(x) = x \ln x + (1 + x) \ln (1 + x)$$

satisfies the functional equation

$$f(x) + (1 + x)f\left(\frac{y}{1 + x}\right) = f(y) + (1 + y)f\left(\frac{x}{1 + y}\right)$$

10. Find a solution of the functional equation in four unknown functions

$$f(x) + (1 + x) g\left(\frac{y}{1 + x}\right) = h(y) + (1 + y)h\left(\frac{x}{1 + y}\right)$$

11. Show that the functional equation

$$f(x, y) + (1 + x)f\left(\frac{u}{1 + x}, \frac{v}{1 + y}\right) = f(u, v) + (1 + u)f\left(\frac{x}{1 + u}, \frac{y}{1 + v}\right)$$

has a solution

$$f(x, y) = A[-x \ln x + (1 + x) \ln (1 + x)] - dx \ln y + d(x + 1) \ln (1 + y).$$

12. Show that $f(x) = (a^x - 1)/c$ satisfies the functional equation

$$f(x + y) = f(x) + f(y) + cf(x)f(y)$$

This functional equation is a non-linear generalisation of (1). By putting $a = (c + 1)^k$ and letting $c \to 0$, deduce the solution of (1).

13. Show that $f(x) = (a^x - b)/c$ satisfies the functional equation

$$f(xy) = b (f(x) + f(y)) + cf(x) f(y) + (b^2 - b)/c$$

Discuss the special cases $b = 0, b = 1$.

14. Consider a non-linear generalisation of (24)

$$f(pq) = f(p) + f(q) + (1 - a) f(p) f(q)$$

Show that its solution which reduces to (26) when $a \to 1$ is given by

$$f(p) = (p^{a-1})/(1 - a)$$

Show that this is also a solution of

$$f(pqr) = f(p) + f(q) + f(r) + (1 - a) \left(f(p) f(q) + f(q) f(r) + f(r) f(p)\right)$$
$$+ (1 - a)^2 f(p) f(q) f(r).$$

15. Show that the average of $f(p)$ of Ex. 14 for probability distribution $(p_1, p_2, \ldots, p_n)$ is

$$\frac{1}{1 - a} \left(\sum_{i=1}^{n} p_i^a - 1 \right)$$

This is called Havrda and Charvat's measure of entropy. Show that this approaches Shannon's measure of entropy (27) as $a \to 1$.

16. Show that $f(x_1, x_2, \ldots, x_n) = x_1^{a_1} x_2^{a_2} \ldots x_m^{a_m}$

$+ (1 - x_1)^{a_1} (1 - x_2)^{a_2} \ldots (1 - x_m)^{a_m}$ satisfies the functional equation

$$f(x_1, x_2, \ldots, x_m) + (1 - x_1)^{a_1} (1 - x_2)^{a_2} \ldots (1 - x_m)^{a_m}$$

$$\times f\left(\frac{y_1}{1 - x_1}, \frac{y_2}{1 - x_2}, \ldots, \frac{y_m}{1 - x_m}\right)$$

$$= f(y_1, y_2, \ldots, y_m) + (1 - y_1)^{a_1} (1 - y_2)^{a_2} \ldots (1 - y_m)^{a_m}$$

$$\times f\left(\frac{x_1}{1 - y_1}, \frac{x_2}{1 - y_2}, \ldots, \frac{x_m}{1 - y_m}\right)$$

Deduce the functional equations satisfied by

$$f(x) = \frac{x^\alpha + (1 - x)^\alpha - 1}{1 - \alpha},$$

$$f(x, y) = \frac{x^\alpha y^{1-\alpha} + (1 - x)^\alpha (1 - y)^{1-\alpha} - 1}{\alpha - 1}$$

$$f(x, y) = \frac{x^\alpha y^{1-\alpha} + (1 - x)^\alpha (1 - y)^{1-\alpha} - x^\alpha - (1 - x)^\alpha}{\alpha - 1}$$

$$g(x) = -x \ln x - (1 - x) \ln (1 - x)$$

$$g(x, y) = x \ln \frac{x}{y} + (1 - x) \ln \frac{1 - x}{1 - y}$$

$$g(x, y) = -x \ln y - (1 - x) \ln (1 - y).$$

8.2 MATHEMATICAL MODELLING THROUGH INTEGRAL EQUATIONS

8.2.1 Integral Equations Arising from a Problem of Elasticity Theory

Consider an elastic beam AB. Let $G(x, \xi)$ denote the displacement at the point x due to unit force applied at ξ. This function $G(x, \xi)$ can be found from the equations of elasticity theory. If the force applied is $f(\xi)d\xi$, then in the linear theory of elasticity, the displacement at x would be $G(x, \xi)$ $f(\xi)d\xi$ and if the force is applied all along the beam, the displacement at the point x would be $\bar{f}(x)$ where

$$\bar{f}(x) = \int_a^b G(x, \xi) f(\xi) \, d\xi \tag{55}$$

Knowing $f(\xi)$ at all points of the beam AB, (55) would enable us to find the displacement $\bar{f}(x)$ of all points of the beam.

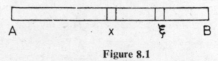

Figure 8.1

The inverse problem is to determine $f(x)$ when $\bar{f}(x)$ is known i.e. to find the force distribution which will cause a desired displacement distribution at all points of the beam. The unknown function in this case is $f(\xi)$ and it occurs under the integral sign. Equations like (55) where the unknown function occurs under the integral sign are called integral equations.

The function $G(x, \xi)$ is called the influence function or the kernel function or Green's function.

Physically, the integral in (55) is arising because the effect of the force applied at different points of the beam is being 'summed up', 'integrated out' and there is 'accumulation' of all effects.

Thus integral equations are likely to arise in physical, biological and social problem where there is 'accumulative effect' in operation.

The function $\bar{f}(x)$ is also called the integral transform of $f(\xi)$ through the kernel $G(x, \xi)$. Inverting this integral transform means finding $f(\xi)$ when $\bar{f}(x)$ is known and this requires the solution of an integral equation.

There are a number of kernels which arise in a large number of applications of mathematics. These give rise to standard integral transforms. Some of these along with the inverse transforms are given in the next section.

8.2.2 Standard Integral Transform Pairs

Laplace Transform pairs

$$\bar{f}(x) = \int_0^\infty e^{-x\xi} f(\xi)d\xi$$

$$f(\xi) = \int_{C-i\infty}^{C+i\infty} e^{-\xi x} \bar{f}(x) \, dx$$

where $C >$ all singularities of $\bar{f}(x)$.
Equation (57) is an integral equation
in the complex plane.

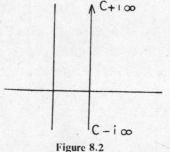

Figure 8.2

Fourier Complex Transform pairs

$$\bar{f}(x) = \int_{-\infty}^{\infty} e^{\pm i\xi x} f(\xi)\, d\xi \tag{58}$$

$$f(\xi) = \frac{1}{2\pi} \int_{-\infty}^{\infty} e^{\mp i\xi x} \bar{f}(x)\, dx \tag{59}$$

Fourier sine (cosine) Transform pairs

$$\bar{f}(x) = \int_{0}^{\infty} \sin(\xi x) f(\xi)\, d\xi \tag{60}$$

$$f(\xi) = \frac{2}{\pi} \int_{0}^{\infty} \sin(\xi x) \bar{f}(x)\, dx \tag{61}$$

$$\bar{f}(x) = \int_{0}^{\infty} \cos(\xi x) f(\xi)\, d\xi \tag{62}$$

$$f(\xi) = \int_{0}^{\infty} \cos(\xi x) \bar{f}(x)\, dx \tag{53}$$

Hilbert Transform Pairs

$$\bar{f}(x) = \frac{1^*}{\pi} \int_{-\infty}^{\infty} \frac{f(\xi)}{x - \xi}\, d\xi \tag{64}$$

$$f(\xi) = \frac{1^*}{\pi} \int_{-\infty}^{\infty} \frac{\bar{f}(x)}{\xi - x}\, dx \tag{65}$$

where * denotes that we are taking the principal value of the integral concerned

$$\int_{-\infty}^{*\ \infty} \frac{f(\xi)}{x - \xi}\, d\xi = \underset{\epsilon \to 0}{Lt} \left[\int_{-\infty}^{x-\epsilon} \frac{f(\xi)}{x - \xi}\, d\xi + \int_{x+\epsilon}^{\infty} \frac{f(\xi)}{x - \xi}\, d\xi \right] \tag{66}$$

If the limits are -1 to 1, the corresponding transform pair is

$$\bar{f}(x) = \frac{1^*}{\pi} \int_{-1}^{1} \frac{f(\xi)}{x - \xi}\, d\xi \tag{67}$$

$$f(\xi) = \frac{1^*}{\pi} \int_{-1}^{1} \sqrt{\frac{1 - x^2}{1 - \xi^2}} \frac{\bar{f}(x)}{\xi - x}\, dx \tag{68}$$

Hankel Transform Pairs

$$\bar{f}(x) = \int_{0}^{\infty} \xi J_n(\xi x) f(\xi)\, d\xi \tag{69}$$

$$f(\xi) = \int_{0}^{\infty} x J_n(\xi x) \bar{f}(x)\, dx \tag{70}$$

Finite Hankel Transform Pairs

$$\bar{f}(x_i) = \int_{0}^{a} f(\xi)\, \xi\, J_n(\xi x_i)\, d\xi \tag{71}$$

$$f(\xi) = \frac{2}{a^2} \sum_{i} \bar{f}(x_i) \frac{J_n(\xi x_i)}{(J_n(a x_i))^2} \tag{72}$$

where x_i is a root of $J_n(ax_i) = 0$ and the summation in (72) is taken over all values of x_i

Mellin Transform Pairs

$$\bar{f}(x) = \int_0^\infty f(\xi)\, \xi^{x-1}\, d\xi \tag{73}$$

$$f(\xi) = \frac{1}{2\pi_i} \int_{c-i\infty}^{c+i\infty} \xi^{-x}\, \bar{f}(x)\, dx \tag{74}$$

Laplace transform pair illustrates the application of these transforms in applied mathematics. A mathematical model is expressed in term of an ordinary linear differential equation. This differential equation is transformed into an algebraic equation by using the Laplace transform. The algebraic equation is solved and the inverse Laplace transform of this solution is obtained and is interpreted as the solution of the original problem. Similarly a partial differential equation with n independent variables is reduced to a partial differential equation with $(n-1)$ independent variables and this leads to a considerable simplification.

Other transforms also reduce the number of independent variables by unity but these are applicable to special types of linear differential equation.

8.2.3 Integral Equations Arising from Differential Equations

Consider the linear differential equation

$$\frac{d^2y}{dx^2} + g_1(x)\, \frac{dy}{dx} + g_2(x)y = F(x) \tag{75}$$

This has in general an infinity of solutions. The solution is however useful if in addition to (75), the boundary conditions

$$y(0) = a, \qquad y'(0) = b \tag{76}$$

are also specified. Here the mathematical model is specified in terms of two parts viz. (i) a differential equation (ii) boundary conditions. Integrating (75), we get

$$\left[\frac{dy}{dx}\right]_0^x + \int_0^x g_1(x)\, \frac{dy}{dx}\, dx + \int_0^x g_2(x)\, dx = \int_0^x F(x)\, dx$$

or $\dfrac{dy}{dx} - b + [yg_1(x)]_0^x - \displaystyle\int_0^x yg_1'(x)\, dx + \int_0^x g_2(x)\, y\, dx = \int_0^x F(x)\, dx$

or $\dfrac{dy}{dx} - b + yg_1(x) - ag_1(0) + \displaystyle\int_0^x [g_2(x) - g_1'(x)]\, y\, dx = \int_0^x F(x)\, dx$

$$\tag{77}$$

Integrating again

$$y - a - bx + \int_0^x [g_1(x) - g_1(0)]\, dx + [x(g_2(x) - g_1'(x))\, y]_0^x$$

$$- \int_0^x x(g_2(x) - g_1'(x))\, y\, dx = \int_0^x dx \int_0^x F(x)\, dx, \tag{78}$$

which is of the form

$$y(x) + \int_0^x y(\xi)\, G(\xi, x)\, d\xi = \varphi(x), \tag{79}$$

where the only unknown function is $y(x)$. This integral equation incorporates the information contained in both the differential equation and the boundary conditions (76) and will in general have a unique solution.

Equation (79) is called Volterra's equation of the second kind. Volterra's equation of the first kind is given by

$$\int_0^x y(\xi)\, G(x, \xi)\, d\xi = \varphi(x) \tag{80}$$

If the limits of the integral are fixed, the corresponding equations are called Fredholm's equation of the first and second kind. These are of the form

$$\int_a^b y(\xi)\, G(x, \xi)\, d\xi = \varphi(x) \tag{81}$$

and

$$y(x) + \int_a^b y(\xi)\, G(x, \xi)\, d\xi = \varphi(x) \tag{82}$$

respectively.

8.2.4 Integral Equations for a Two-Point Boundary Value Problem

We consider the two-points boundary value problem

$$y'' = f(x), \qquad y(0) = 0, \qquad y(b) = 0, \tag{83}$$

where one boundary condition is specified at $x = 0$ and the other is specified at $x = b$. In the last subsection, both boundary conditions were specified at $x = 0$.

We can write the differential equation and boundary conditions as

$$y''(x) = \int_{-\infty}^{\infty} f(\xi)\, \delta(x - \xi)\, d\xi, \quad y(0) = 0, \quad y(b) = 0 \tag{84}$$

where $\delta(x - \xi)$ is Dirac's delta function which vanishes when $x > \xi$ and when $x < \xi$ and takes an infinite value at $x = \xi$ in such a way that

$$\int_{-\infty}^{\infty} \delta(x - \xi)\, dx = 1 \tag{85}$$

As such we first consider the boundary value problem

$$y''(x) = \delta(x - \xi), \quad y(0) = 0, \quad y(b) = 0 \tag{86}$$

This means that

$$y''(x) = 0; \quad 0 < x < \xi; \quad y''(x) = 0, \quad \xi < x < b \tag{87}$$

giving solutions

$$y = ax + b, \quad 0 < x < \xi; \quad y = cx + d, \quad \xi < x < b \tag{88}$$

Since $y(0) = 0$, $y(b) = 0$, (88) gives

$$y = ax, \quad 0 < x < \xi, \quad y = c(x - b), \quad \xi < x < b \tag{89}$$

There are two constant viz. a and c yet to be determined. For determining these, we use the two following conditions viz.

(i) $y(x)$ is continuous at $x = \xi$ i.e. $y(\xi + 0) = y(\xi - 0)$ $\qquad$ (90)

(ii) From (86)

$$(y'(x))_{\xi-0}^{\xi+0} = \int_{\xi-0}^{\xi+0} \delta(x - \xi)\, dx = \int_{-\infty}^{\infty} \delta(x - \xi)\, dx = 1 \tag{91}$$

i.e. the derivative $y'(x)$ is discontinuous at ξ and the jump in its value is unity. From (87), (90) and (91)

$$a\xi = c\xi - b, \quad c - a = 1 \tag{92}$$

so that the solution of (88) is

$$y = G(x, \xi), \tag{93}$$

where

$$G(x, \xi) = \frac{\xi - b}{b}\, x, \quad 0 \leqslant x \leqslant \xi$$

$$= \frac{\xi}{b}\, (x - b), \quad \xi \leqslant x \leqslant b \tag{94}$$

The graph of $G(x, \xi)$ is shown in Figure 8.3. It shows that the Green's function is continuous at all points between 0 and b, but its derivative does not exist at $x = \xi$.

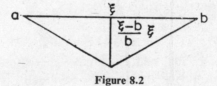

Figure 8.2

By using the superposition principle, the solution of (83) and (84) is given by

$$y(x) = \int_0^b G(x, \xi)\, f(\xi)\, d\xi \tag{95}$$

This is not an integral equation, it is the solution of the boundary value problem (83). However consider the more general boundary value problem

$$y'' - w^2 y = f(x), \quad y(0) = 0, \quad y(b) = 0 \tag{96}$$

and consider a corresponding integral equation

$$y(x) - w^2 \int_0^b G(x, \xi)\, y(\xi)\, d\xi = \int_0^b G(x, \xi)\, f(\xi)\, d\xi \tag{97}$$

We investigate whether the solution of (97) satisfies the differential equation and boundary conditions of (96).

Since from (94), $\qquad G(0, \xi) = 0, \qquad G(b, \xi) = 0,$ \hfill (98)

(97) gives $\qquad\qquad\qquad y(0) = 0, \qquad y(b) = 0$ \hfill (99)

and so (97) satisfies the boundary conditions (99). Now differentiating (97) twice, we get differential equation in (96).

Thus the solution of the two-points boundary value problem (96) is reduced to the solution of the integral equation (97).

8.2.5 A More General Two-Point Boundary Value Problem

Consider the boundary value problem

$$y'' + g_1(x)y' + g_2(x) = f(x); \ y(0) = 0, \ y(b) = 0, \tag{100}$$

Let the solution of

$$y'' + g_1(x)y' + g_2(x) = 0 \tag{101}$$

be $\qquad\qquad\qquad y = A_1 y_1(x) + A_2 y_2(x), \tag{102}$

then the solution of

$$y'' + g_1(x)y' + g_2(x)y = \delta(x - \xi), \ y(0) = 0, \ y(b) = 0 \tag{103}$$

is given by

$$y = c_1 y_1(x) + c_2 y_2(x), \qquad 0 \leqslant x \leqslant \xi \tag{104}$$

$$y = d_1 y_1(x) + d_2 y_2(x) \qquad \xi \leqslant x \leqslant b, \tag{105}$$

where the constants c_1, c_2, d_1, d_2 are obtained from the equations

$$c_1 y_1(0) + c_2 y_2(0) = 0, \qquad d_1 y_1(b) + d_2 y_2(b) = 0 \tag{106}$$

$$c_1 y_1(\xi) + c_2 y_2(\xi) = d_1 y_1(\xi) + d_2 y_2(\xi) \tag{107}$$

and $\qquad c_1 y_1'(\xi) + c_2 y_2'(\xi) - d_1 y_1'(\xi) - d_2 y_2'(\xi) = 1 \tag{108}$

Knowing c_1, c_2, d_1, d_2, (104) and (105) determine the Green's function $G(x, \xi)$ for the present problem and then the solution of (100) is

$$y(x) = \int_0^b G(x, \xi) f(\xi) \, d\xi \tag{109}$$

8.2.6 Integral Equations in Population Dynamics

Knowing

$\qquad {}_x p_0 =$ probability of a female of age zero, i.e. a female just born, surviving till age x

$\qquad {}_t p_{x-t} =$ probability of a female of age $x - t$ surviving till age $x (x \geqslant t)$

$\qquad \lambda(x) \Delta t =$ average number of births to a female with age between x and $x + \Delta x$

$\qquad F(x, 0) =$ initial number of females of age x at time $t = 0$, it is required to find

$\qquad F(x, t) \Delta x =$ number of females at time t of ages between x and $x + \Delta x$

$\qquad B(t) \Delta t =$ number of total female births in time interval $t, t + \Delta t$

The above definitions lead to the following relations:

(1) $$F(x, t) = B(t - x) \,_xp_0; \; x \leqslant t \qquad (110)$$

This follows since $B(t - x)$ denotes the number of females born at time $t - x$ and $_xp_0$ gives the probability of their surviving for x years to become of age x at time t. Thus (110) expresses the fact that the number of females of age x at time is equal to the number of females born at time $t - x$ who have survived for x years.

(ii) $$F(x, t) = F(x - t, 0)\,_tp_{x-t} \; x \geqslant t \qquad (111)$$

This expresses the fact that the number of females of age x at time t is equal to the number of females of age $x - t$ at time 0 who have survived for t years to become of x years.

(iii) $$B(t)\varDelta t = \int_\alpha^\beta F(x, t)\lambda(x) \, dx \, \varDelta t, \qquad (112)$$

where (α, β) gives the reproductive age-group interval so that

$$\lambda(x) = 0 \text{ when } x < \alpha \text{ and when } x > \beta \qquad (113)$$

Equation (112) expresses the fact that the total number of female birth taking place during time interval $(t, t + \varDelta t)$ is obtained by summing or integrating the number of female births due to females of all ages in the reproductive age-group. In view of (113), equation (112) can also be written as

$$B(t) = \int_0^\infty F(x, t)\lambda(x) \, dx$$
$$= \int_0^t F(x, t)\lambda(x) \, dx + \int_t^\infty F(x, t)\lambda(x) \, dx \qquad (114)$$

Now using (110) and (111), we get

$$B(t) = \int_0^t B(t - x)\,_xp_0\lambda(x) \, dx + \int_t^\infty F(x - t, 0)\,_tp_{x-t}\lambda(x) \, dx$$
$$= \int_0^t B(t - x)\,_xp_0\lambda(x) \, dx + \int_0^\infty F(u, 0)\,_tp_u\lambda(t + u) \, du \qquad (115)$$

Now $_xp_0$ and $_tp_u$ can be found by statistical analysis of census data, $F(u, 0)$ is the number of females of age u at time $t = 0$ and is supposed to be given. The birth rate $\lambda(x)$ for all age-groups is also given. As such the only unknown function in (115) is $B(t)$. Thus (115) gives Volterra integral equation of the second kind to determine $B(t)$. Knowing $B(t)$, equation (110) would enable us to know the female population of all ages $x \leqslant t$. The female population of ages $\geqslant t$ can be determined from equation (111).

It may be noted that if $t \geqslant \beta$, the second integral on the RHS (114) or (115) vanishes and our integral equation becomes

$$B(t) = \int_0^\infty B(t - x)\,_xp_0\lambda(x) \, dx, \qquad t \geqslant \beta, \qquad (116)$$

which is a linear homogeneous integral equation.

8.2.7 Mathematical Modelling Through Integral Equations

An integral equation mathematical model can arise whenever the effect of an unknown function is summed over or integrated over a period of time or an interval in space. The total effect of a function $f(z)$ through an influence or kernel function $G(x, z)$ is given by

$$\int_0^\infty f(z)\, G(x, z)\, dz \quad \text{or} \quad \int_0^x f(z)G(x, z)\, dz \qquad (117)$$

If some physical or biological laws enable us to express the effect in terms of $f(x)$, or in terms of a known function, we get an integral equation.

Thus in (58), the total influence of an unknown force distributed over the length of a beam is expressed in terms of a known displacement function $f(x)$ and in (115), the number of births at time t is expressed in terms of the number of births at all earlier times.

Similarly in environment studies, pollution effects in air or water are cumulative, in economic studies, the effects of economic policies are cumulative, in elastic substances with memory, displacements accumulate and in all these cases mathematical models are in terms of integral equations.

A differential equation models a local situation. If a differential equation holds at all points of an interval or region, it can model a global situation, provided in addition, boundary conditions are also specified. We have seen above that a differential equation along with its boundary conditions is equivalent to an integral equation. Thus an integral equation models a global situation. In principle, a situation which can be modelled through differential equations and boundary conditions should be capable of being modelled through integral equations.

Some interesting examples of mathematical modelling in physics through integral equations are given in Morse-Feshbch 'Methods of Theoretical Physics' Chapters 8 and 11. These include ths following:

(i) *Transport Theory*. The integral here arises because particles can have momentum value $\vec{P}$ after collision when its initial momentum could have any value $\vec{P}_0$ and as such we have to integrate over all possible value of $\vec{P}_0$.

(ii) *Accoustic Theory*. Here the behaviour of a membrane at a point depends on the behaviour at all point of the membrane and the relationship is expressed by an integral.

(iii) *Radiation Theory*. Radiation is transmitted through all points of a medium and the effects at all points have to be summed up.

(iv) *Wave Mechanics*. Here Shrodinger equation in differential equation form is transformed to an integral equation form

(v) *Helmholtz Equation*. Here we reduce a differential equation in two dimensional space to an integral equation in one dimension, thus leading to considerable simplification.

EXERCISE 8.2

1. Show that the solution of differential equations $y' = f(x, y)$ subject to $y(x_0) = y_0$ is the same as that of the integral equation

$$y(x) = \int_{x_0}^{x} f(x, y) \, dx + y_0.$$

2. Show that the solution of the differential equation $y'' = f(x, y)$ subject to $y(x_0) = y_0$, $y'(x_0) = y_0'$ is the same as that of the integral equation

$$y(x) = \int_{x_0}^{x} (x - z) f(z, y(z)) \, dz + y_0 + y_0' (x - x_0)$$

3. Show that the general solution of $y'' = f(x, y)$ is the solution of the integral equation

$$y(x) = \int_{0}^{x} (x - z) f(z, y(z)) \, dz + c_1 + c_2 x,$$

where c_1, c_2 are arbitrary constants.

4. In Ex. 3, determine c_1, c_2 so that $y(a) = a$, $y(b) = c$

5. Solve problems similar to those of Ex. 3 and Ex. 4 for the differential equation

$$y'' + p(x)y' = w(x)$$

6. Consider the partial differential equation of forced transversed vibrations of a string fixed at the end points $x = 0$, $x = b$ viz.

$$\rho u_{tt} = \tau u_{xx} + f(x, t)$$

Substitute $f(x, t) = \varphi(x) \cos wt$, $u(x, t) = v(x) \cos wt$, to get at the boundary value problem

$$\tau v'' = -\rho w^2 v - \varphi(x) \, ((0 \leqslant x \leqslant b), \, v(0) = 0, \, v(b) = 0,$$

Express this in terms of an integral equation both when $w = 0$ and $w \neq 0$.

7. Solve equation (116) (see page 120 of J.N. Kapur, Mathematical Models in Biology and Medicine).

8. Let L^{-1} denote inverse Laplace transform, show that if

$$L^{-1}(f(s)) = F(t), \text{ then } L^{-1}[f^{(n)}(s)] = (-1)^n t^n F(t)$$

and

$$L^{-1}\left[\frac{f(s)}{s}\right] = \int_{0}^{t} F(u) \, du.$$

9. Show that the Fourier transform of $F(t)$ which is equal to $e^{-xt}\varphi(t)$ when $t > 0$ and is zero when $t < 0$, is the Laplace transform of $\varphi(t)$ i.e. it is equal to $\int_{0}^{\infty} e^{-zt}\varphi(t) \, dt$, when $z = x + iy$.

10. Let $\bar{f}_n(s)$ be the Hankel transform of order n of the function $f(x)$ and $\bar{f}_n'(s)$ be transform of $f'(x)$, then show that

$$\bar{f}_n'(s) = -\frac{s}{2n} [(n + 1) f_{n-1}'(s) - (n - 1) \bar{f}_{n+1}'(s)].$$

11. Develop a model for determining the age structure of the trees of a forest at any time t when the initial number of trees of age $x \mp \frac{1}{2} dx$ $(0 < x < b)$ is given, when the number of new plants planted in time $t \mp \frac{1}{2} dt$ is given and when the number of trees of age group x harvested in time $t \mp \frac{1}{2} dt$ is also given. It is given that no tree of age $< a$ is cut and all trees of age $\geqslant b$ are cut.

12. Develop a model for finding the rate of growth of pollution when the number of factories in an area increases linearly and the effect of the pollutant produced by a factory decreases with time and a certain percentage of pollutants is being constantly destroyed by an antipollution agency

8.3 MATHEMATICAL MODELLING THROUGH DELAY-DIFFERENTIAL AND DIFFERENTIAL-DIFFERENCE EQUATIONS

8.3.1 Single Species Population Models

Let $N(t)$ be the population size at time t and let $B(t)$ and $D(t)$ be the birth and death-rates i.e. the numbers of persons being born or dying per unit individual per unit time, then we get the model

$$\frac{dN}{dt} = B(t)N(t) - D(t)N(t) \tag{118}$$

Now the birth-rate at time t depends on the population size at time $t - \tau$ and the death-rate depends on the population size at time t. The simplest assumptions are

$$B(t) = b_1 - b_2 N(t - \tau), \ D(t) = d_1 + d_2 N(t); \ b_1, b_2, d_1, d_2 > 0 \tag{119}$$

From (118) and (119)

$$\frac{dN}{dt} = (b_1 - d_1)N(t) - b_2 N(t)N(t - \tau) - d_2 N^2(t) \tag{120}$$

This is called a Delay-Differential equation since it includes both a differentiation operator and a delay effect term. If there is no delay, it reduces to an ordinary differential equation

8.3.2 Prey-Predator Model

Let $N_1(t)$ and $N_2(t)$ be the populations of the prey and predator species respectively, then assuming that the contacts between prey and predator species result in instantaneous loss to the prey species, but a delayed gain to the predator species, we get the model

$$\frac{dN_1}{dt} = aN_1(t) - bN_1(t)N_2(t),$$

$$\frac{dN_2}{dt} = -pN_2(t) + qN_2(t)N_1(t - 1) \tag{121}$$

We may also replace the second equation of (121) by

$$\frac{dN_2}{dt} = -pN_2(t) + qN_2(t - 1)N_1(t - 1) \tag{122}$$

We may also consider the more general model

$$\frac{dN_1}{dt} = aN_1(t) - aN_1^2(t) - bN_1(t)N_2(t),$$

$$\frac{dN_2}{dt} = -pN_2(t) - rN_2^2(t) + qN_2(t)N_1(t - 1) \tag{123}$$

We may also consider the symmetric model

$$\frac{dN_1}{dt} = aN_1(t) - bN_1(t)N_2(t - \tau),$$

$$\frac{dN_2}{dt} = -pN_2(t) + qN_2(t)N_1(t - \tau) \tag{124}$$

8.3.3 Multispecies Model

Consider

$$\frac{dN_i}{dt} = a_iN_i(t) + \sum_{j=1}^{n} b_{ij}N_i(t)N_j(t - \tau_{ij}), \ i = 1, 2, \ldots n \tag{125}$$

These models in terms of systems of delay-differential equation can include both prey-predator and competition interactions.

8.3.4 Stability of Equilibrium Positions

For solving (120), knowledge of $N(0)$ is not enough, we have to know the values of $N(t)$ from time $-\tau$ to 0. Even after knowing all these values, the solution of (120) is not easy.

We can however find the equilibrium or steady-state solutions by putting $dN/dt = 0$ and replacing $N(t)$ and $N(t - \tau)$ by the equilibrium population size $\overline{N}$, thus getting

$$\overline{N} = (b_1 - d_1)/(b_2 + d_2) \tag{126}$$

To discuss the stability of this position, we substitute $N(t) = \overline{N} + u(t)$ in (120) to get

$$\frac{du}{dt} = (b_1 - d_1)(\overline{N} + u(t)) - b_2(\overline{N} + u(t))(\overline{N} + u(t - \tau))$$

$$-d_2(\overline{N} + u(t))^2 \tag{127}$$

Neglecting squares and products of $u(t)$, $u(t - \tau)$ and using (126) we get the linear delay-differential equation

$$\frac{du}{dt} = (b_1 - d_1)u(t) - b_2\overline{N}(u(t) + u(t - \tau)) - 2d_2\overline{N}u(t) \tag{128}$$

Trying the solution

$$u(t) = Ae^{-\lambda t}, \tag{129}$$

ve get

$$\lambda = (b_1 - d_1) - b_2\overline{N}(1 + a^{-\lambda\tau}) - 2d_2\overline{N} \tag{130}$$

This is an equation to solve for λ, which involves both algebraic and non-algebraic (exponential) functions of λ. If all of its roots have negative real parts, the equilibrium position is stable.

We may substitute $\lambda = r + is$ in (130) and equate real and imaginary parts of both sides to get two equations in r and s. By eliminating s between these two equations, we can get a single equation to determine r. If all the roots of this equation are negative real numbers, the equilibrium position is stable.

The same method can be applied to discuss the stability of all equilibrium positions of all delay-differential equation models.

8.3.5 A Model for Growth of Population Inhibited by Cumulative Effects of Pollution

The rate of growth of a population at any time is inhibited due to the metabolic products produced by populations at all earlier times. The effect of pollution produced by the population between time $t - \tau$ and $t - \tau - d\tau$ on the rate of growth of population at time t may be

$$-ck(\tau)N(t)N(t - \tau)\,d\tau, \tag{132}$$

where $k(\tau)$ is a decreasing function of τ, so that our mathematical model is

$$\frac{dN}{dt} = aN(t) - cN(t)\int_0^\infty k(\tau)N(t - \tau)\,d\tau, \tag{133}$$

which can also be written as

$$\frac{dN}{dt} = aN(t) - cN(t)\int_{-\infty}^t k(t - u)N(u)\,du \tag{134}$$

Equations (133) and (134) are integro-differential equations.

Models in terms of integro-differential equations arise when our physical principles involves both a rate of change of some function as well as the sum or integral or cumulative effects on that function.

8.3.6 Prey-Predator Model in Terms of Integro-Differential Equations

If $N_1(t)$, $N_2(t)$ are the populations of the prey and predator species at time t and the interaction effects are cumulative over time, we get the model

$$\frac{dN_1}{dt} = aN_1(t) - bN_1(t)\int_0^\infty k_1(\tau)N_2(t - \tau)\,d\tau \tag{135}$$

$$\frac{dN_2}{dt} = -pN_2(t) + qN_2(t)\int_0^\infty k_2(\tau)N_1(t - \tau)\,d\tau \tag{136}$$

The kernel functions are usually monotonic decreasing functions of τ which can always be normalised to give

$$\int_0^\infty k_1(\tau)\,d\tau = 1, \quad \int_0^\infty k_2(\tau)\,d\tau = 1. \tag{137}$$

8.3.7 Stability of the Prey-Predator Model

Putting $N_1(t) = \overline{N_1}$, $N_2(t) = \overline{N_2}$ in (135), (136), we get, in view of (137)

$$\overline{N_1} = \frac{p}{q}, \quad \overline{N_2} = \frac{a}{b} \tag{138}$$

Substituting

$$N_1(t) = \frac{p}{q} + u_1(t), \quad N_2(t) = \frac{a}{b} + u_2(t) \tag{139}$$

in (135), (136) and neglecting squares and products of $u_1(t)$, $u_2(t)$, we get the linear equations

$$\frac{du_1}{dt} = -\frac{bp}{q} \int_0^\infty u_2(t - \tau)k_1(\tau)\, d\tau \tag{140}$$

$$\frac{du_2}{dt} = \frac{aq}{b} \int_0^\infty u_1(t - \tau)k_2(\tau)\, d\tau \tag{141}$$

Substituting

$$u_1(t) = A_1 e^{\lambda t}, \qquad u_2(t) = A_2 e^{\lambda t}, \tag{142}$$

we get

$$A_1\lambda + \frac{bp}{q} A_2 k_1^*(\lambda) = 0, \quad A_2\lambda - \frac{aq}{b} A_1 k_2^*(\lambda) = 0, \tag{143}$$

where $k_1^*(\lambda)$, $k_2^*(\lambda)$ are Laplace transforms of $k_1(t)$, $k_2(t)$ respectively

Eliminating $\qquad\qquad A_1/A_2$, we get

$$\lambda^2 + ab\, k_1^*(\lambda)k_2^*(\lambda) = 0 \tag{144}$$

The equilibrium position given by (138) would be stable if the real parts of all the roots of (144) are negative.

8.3.8 Special Cases

(a) If $k(\tau) = \delta(\tau)$ where $\delta(\tau)$ is Dirac's delta function satisfying

$$\int_0^\infty \delta(\tau)\, d\tau = 1, \quad \int_0^\infty f(\tau)\delta(\tau)\, d\tau = f(0), \quad \int_0^\infty f(t - \tau)\delta(\tau)\, d\tau = f(t), \tag{145}$$

then (133) reduces to

$$\frac{dN}{dt} = aN(t) - cN^2(t), \tag{146}$$

so that the ordinary differential equation model is a special case of integro-differential equation model.

(b) If $k(\tau) = \delta(\tau - \tau_0)$, then (133) and (145) give

$$\frac{dN}{dt} = aN(t) - eN(t)N(t - \tau_0) \tag{147}$$

so that the delay-differential equation model is also a special case of the integro-differential equation model.

(c) If $k_1(\tau) = k_2(\tau) = \delta(\tau)$ then (135) and (136) give the prey-predator model in terms of ordinary differential equations and the characteristic equation (144) becomes

$$\lambda^2 + ab = 0. \tag{148}$$

(d) If

$$k_i(\tau) = \frac{\exp(-\alpha_i\tau)\alpha_i^{ni} n_i^{\tau-1}}{\Gamma(n_i)}, \qquad i = 1, 2 \tag{149}$$

then

$$k_i{}^*(\lambda) = \int_0^\infty \frac{\exp(-\alpha_i\tau)n_i^{\tau-1}\alpha_i^{ni}e^{-\lambda\tau}}{\Gamma(n_i)}\, d\tau = \left(\frac{\alpha_i}{\alpha_i + \lambda}\right)^{ni} \tag{150}$$

and (144) becomes

$$\lambda^2 + ab\left(\frac{\alpha_1}{\alpha_1 + \lambda}\right)^{n_1}\left(\frac{\alpha_2}{\alpha_2 + \lambda}\right)^{n_2} = 0, \tag{151}$$

which is an algebraic equation of degree $n_1 + n_2$ and the equilibrium position will be stable if the real parts of all its roots are negative.

8.3.9 Differential-Differences Equations Models in Relation to Other Models

In section 6.5 we modelled birth-death and epidemic models in terms of systems of differential-difference equations, while in section 8.1 we have modelled population-growth and prey-predator models in terms of delay-differential equations. In both types of equations, both differencing and differentiating operators are involved, but while in differential-difference equations, these differentiating and difference operators are with respect to different independent variables, in delay-differential equations, these are with respect to the same independent variable.

Thus in birth-death and epidemic models, there were two independent variables since probabilities there depended both on the number of individuals in the system, which is a discrete variate, and the time which is a continuous variable. Here we did differencing with respect to number of individuals and we did differentiation with respect to time and we got differential-difference equations.

In population models, $N(t)$ is essentially a discrete variable which is a function of a continuous variable i.e. of time t. However we idealise $N(t)$ as a continuous dependent variable which is a function of the continuous variable time. We essentially use the operator of differentiation, but we also introduce the operation of differencing by introducing a delay effect and we get a delay-differential equation model.

If the delay factor is continuously distributed, we get an integro-differential equation.

What we have called as a system of differential-difference equations may also be called a system of differential equations also, depending on the point of view we take. Thus if in birth-death processes, we take probability as a dependent variate depending on two independent variates, we call the result-ing systems of equations as a system of differential-difference equations for determining the single dependent variable. We may however regard

$$p_0(t), p_1(t), p_2(t), \ldots, p_n(t), \ldots$$

as a set of infinite dependent variables and then we may call the system of equations as a system of infinite differential equation to determine an infinite number of dependent variables, all depending on time t.

Similarly in age-structured population models, we may consider (3.21) as a system of differential equations to determine n different variables

$$x_1(t), x_2(t) \ldots x_n(t)$$

all depending on time t or we may regard (3.21) as a system of differential-difference equations to determine the single dependent variable $x_n(t)$ viz. the population which is a function of n, corresponding to the age-group and t corresponding to time.

EXERCISE 8.8

1. For the differential and delay-differential equations models given below, verify the given equilibrium positions and characteristic equations

(i) $dN/dt = bN - dN^2$; $\bar{N} = b/d$, $\lambda + b = 0$

(ii) $dN/dt = bN(t) - dN(t)N(t - 1)$; $\bar{N} = b/d$; $\lambda + be^{-\lambda} = 0$

(iii) $dN/dt = bN(t) - d_1N^2(t) - d_2N(t)N(t - 1)$

   $\bar{N} = b/(d_1 + d_2)$; $\lambda + K_1e^{-\lambda} + K_2 = 0$

(iv) $dN/dt = (b - d_1N(t - 1) - d_2N(t - 2))N(t)$

   $\bar{N} = b/(d_1 + d_2)$, $\lambda + K_1e^{-\lambda} + K_2e^{-2\lambda} = 0$

(v) $dN/dt = (b - d_0N(t) - d_1N(t - 1) - \ldots - d_mN(t - m))N(t)$

   $$\bar{N} = \frac{b}{d_1 + d_2 + \ldots + d_m};$$

   $\lambda + d_0 + K_1e^{-\lambda} + K_2e^{-2\lambda} + \ldots + K_me^{-m\lambda} = 0.$

2. For the following integro-differential equation models, verify the given equilibrium positions and the characteristic equations

(i) $dN/dt = b - d\displaystyle\int_{-\infty}^{t} N(t)k(t - s)\,ds$

   $\bar{N} = b/d$, $\lambda + bk^*(\lambda) = 0$

(ii) $dN/dt = a(N) - bN^2 - dN\displaystyle\int_{0}^{\infty} N(t - s)k(s)\,ds$

   $\bar{N} = \dfrac{a}{b + d}$, $\lambda + \dfrac{a}{b + d}(b + dk^*(\lambda)) = 0$

(iii) $dN/dt = aN - bN^2 - cp\left(\int_0^\infty N(t - s)k(s)\ ds\right)^2$

$$\bar{N} = \frac{a}{b + cp}, \qquad \lambda°(b + c) + ab - ccp - 2acp\ k*(\lambda) = 0$$

(iv) $dN/dt = bN - dN\int_0^\infty N(t - s)k(s)\ ds, \qquad m > 1$

$$\bar{N} = \frac{b}{d}, \qquad \lambda = b - d\left(\frac{b}{d}\right)^{m/(-1)}$$

3. Interpret each of the models in Exercises 1 and 2 as population growth models.

4. Obtain special cases of models in Ex. 2 when the kernel functions is a Dirac delta function or the distributed delay function given by (144).

5. Form ten deterministic models in terms of systems of ordinary differential equation, find the corresponding stochastic models and the system of differential-difference equations for probabilities. Find in each case the differential equations for the probability generating function and discuss whether you have in each case sufficient initial and boundary conditions to solve the differential equation that arises.

9

Mathematical Modelling Through Calculus of Variations and Dynamic Programming

9.1 OPTIMISATION PRINCIPLES AND TECHNIQUES

9.1.1 Mathematical Models for Description, Prediction, Optimization and Control

In the preceding eight chapters, we have obtained mathematical models in terms of ordinary and partial differential equations, difference equations, functional equations, delay-differential, integral and integro-differential equations and graph theory concepts by using the principles like those of continuity, conservation of mass, momentum and energy, Newton's laws of motion, plausible relations between growth, birth and death rates and between supply, demand, price, national income, savings, investments etc.

Most of these models can describe situations precisely and quantitatively, these can also predict what will happen at some future time and at some specified point of space.

However one of the main goals of mathematical modelling is optimization of some objective function subject to some given constraints.

Thus in economics, one may like to maximize profits or production and minimize losses, subject to given constraints on resources. In space flight, one may like to minimize the time of flight or consumption of energy, a government may like to give maximum welfare to its citizens subject to given financial and human resources and so on.

One can maximize or minimize only if one has some variables under one's control whose values one can choose to get desired optimization. Thus the problems of control are intimately related to problems of optimization.

9.1.2 Some Optimization Principles

Throughout history and in particular during the last three centuries, scientists and engineers have tried to identify those entities by maximizing or minimizing which, they can get the various laws of science and social behaviour. This search has led to many basic optimizing principles. We state some of these below:

Principle of Minimum Potential Energy: In stable equilibrium, the potential energy of a mechanical system is *least*

Fermat's Principle of Least Time: Light travels from one point to another in such a way as to take *least* possible time. From this principle, we can deduce all the laws of optics.

Hamilton's Principle and Extended Hamilton's Principle: Under certain conditions, the actual motion of a dynamical system is found by obtaining the *extremal* of the Hamiltonian integral $\int_{t_0}^{t} (T - V) \, dt$, where T and V are kinetic and potential energies of the system.

Principle of Least Action: Here the actual motion is found by finding the *extremum* of the action integral $\int_{t_1}^{t_2} T \, dt$.

Principle of Maximum Likelihood: The best estimate of a parameter θ, given a random sample $x_1, x_2, \ldots, x_n$ from a population with density function $f(x, \theta)$ is obtained by *maximizing* the likelihood function

$$L(x_1, x_2, \ldots, x_n; \theta) = f(x_1, \theta) f(x_2, \theta) \ldots f(x_n, \theta). \tag{1}$$

Principle of Least Squares: The best estimates for a, b for fitting the straight line

$$y = a + bx \tag{2}$$

to the data points (x_i, y_i), $(i = 1, 2, \ldots, n)$ are obtained by *minimizing*

$$\sum_{i=1}^{n} (y_i - a - bx_i)^2 \tag{3}$$

Principle of Minimum Chi Square: Let o_i be the observed frequency in the ith class and let e_i be the expected frequency in this class based on the hypothesis that the population parameter is θ, then choose θ so as to minimize

$$\sum_{i=1}^{n} \frac{(o_i - e_i)^2}{e_i} \quad \text{or} \quad \sum_{i=1}^{n} \frac{(o_i - e_i)^2}{o_i} \tag{4}$$

Principle of Minimum Expected Number of Observations in Sequential Analysis: In testing of hypotheses, we either keep the error of first kind and number of observations fixed and try to seek to *minimize* the error of the second kind or keep the errors of both kinds at fixed levels and seek to *minimize* the expected number of observations. Here error of first (second) kind arises when we reject (accept) a hypothesis which is true (false).

Principle of Optimal Design of Experiments: We seek to design experiments which give the *maximum* possible information.

Principle of Choosing Optimal Algorithms: We seek those algorithms which *minimize* the time taken or *maximize* the reliability of the results.

Pareto Optimality Principle: We seek to find all those economic situations which are *jointly optimal* i.e. which are such that any deviation from a situation for the benefit of any individual can only be at the cost of some other individuals.

Principle of Optimal Choice of Strategies by Players: Here each player wants to choose a strategy which *maximizes* his expected gain, whatever be the strategy employed by the other players.

Principle of Choice of Optimal Aeronautical Shapes: Here we choose shapes which give *maximum* lift and *minimum* drag.

Principle of Design of Optimal Structures: Here we seek a structure with *maximum* loads and *minimum* costs.

Principle of Optimal Reliability: Here we seek to choose systems which *maximize* reliability at given cost or *minimise* cost for given reliability.

Principle of Choice of Optimal Decisions: If there is uncertainty, we seek a decision which *maximizes* expected utility.

Principle of Optimal Choice of Portfolios: Here we seek to find portfolios which *maximize* expected return and *minimize* variance.

Principle of Optimal Feature Extraction in Pattern Recognition: Here we seek features which result in *minimum* loss of information or in *minimum* loss of power of discrimination or which lead to *minimum* variability within classes and *maximum* variability between classes or which lead to *minimum* interdependence of components of feature vector.

Principle of Maximum Entropy: Here we seek a probability distribution which has *maximum* entropy or uncertainty out of all those distributions which have prescribed moments.

Principle of Minimum Discrimination Information: Here we seek a distribution which has *minimum* directed divergence from a given distribution, out of all those that have prescribed moments.

Principle of Optimality: Here the *best n*-stage policy is obtained by combining the result of an arbitrary one-stage policy decision with the best remaining $(n-1)$-policy decision and then choosing the first decision to optimize the result.

Optimization Models arise as a result of application of one or more of these principles or even otherwise. New and challenging mathematical problems arise in obtaining the results from those optimizing models. A number of classical as well as new techniques are available for solving these problems and even there new problems motivate the developments of new techniques almost everyday.

9.1.3 Some Techniques for Optimization

(i) *Method of Inequalities*: If it can be shown that

$$f(x_1, x_2, \ldots, x_n) \geqslant A \quad \text{or} \quad \leqslant B \tag{5}$$

whenever $\qquad g_i(x_1, x_2, \ldots, x_n) = b_i \qquad i = 1, 2, \ldots, m \qquad (6)$

and for some $x_1, x_2, \ldots, x_n, f(x_1, x_2, \ldots, x_n) = A$ or B, then
A gives the minimum value of the function and B is the maximum value of
the function subject to constraints (6).

(ii) *Method of Differential Calculus:* The function $f(x_1, x_2, \ldots, x_n)$ has a
local maximum at the point $x_{10}, x_{20}, \ldots, x_{n0}$ if all the first order partial
derivatives vanish at this point and if the matrix of all the second order
partial derivatives at the point is negative definite. Lagrange's method is
used when the function is to be maximized subject to some constraints.

(iii) *Method of Calculus of Variations*: In differential calculus, we have
to find $x_1, x_2, \ldots, x_n$ for which $f(x_1, x_2, \ldots, x_n)$ is maximum or minimum.
In calculus of variations, we have to find functions

$$u_1(x_1, x_2, \ldots, x_m), \quad u_2(x_1, x_2, \ldots, x_m), \quad \ldots, \quad u_n(x_1, x_2, \ldots, x_m)$$

for which a function of these functions is maximum or minimum. We shall
study some mathematical modelling through this technique in section 9.2.

(iv) *Method of Dynamic Programming:* This is useful for multistage decision making and for optimizing functions of several variables. Mathematical
Modelling through this technique will be discussed in section 9.3.

(v) *Method Based on Maximum Principle:* This is useful for control
problems. Mathematical Modelling through this technique will be discussed
in section 10.4.

(vi) *Mathematical Programming Techniques*: These are special techniques
developed for optimizing a function $f(x_1, x_2, \ldots, x_n)$ subject to constraints
$g_r(x_1, x_2, \ldots, x_n) \leqslant a_r (r = 1, 2, \ldots, m)$ and non-negatively constraints
$x_1 \geqslant 0, x_2 \geqslant 0, \ldots, x_n \geqslant 0$. If the functions are linear, we need technique
of *linear programming*, if the variables are required to be integers, we need
special technique of *integer programming*. We also have special techniques
for *quadratic programming, non-linear fractional programming, convex programming, stochastic programming* etc. Mathematical Modelling through
these techniques will be discussed in sections 10.1 and 10.2.

EXERCISE 9.1

1. A right circular cone can be placed on a table in different positions.
Show that it is in stable equilibrium when its potential energy is minimum.

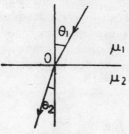

2. Light travels from a given point
A in one medium with refractive
index μ_1 to another given point B in
another medium with refractive index
μ_2. The velocities of light in the two
media are c/μ_1 and c/μ_2 respectively.
Show that the time taken is least
when 0 is so chosen that

Figure 9.1

$$\mu_1 \sin \theta_1 = \mu_2 \sin \theta_2.$$

3. The density function for a population is $1/\sqrt{2\pi} \exp\left(-1/2(x - m)^2\right)$. Use the principle of maximum likelihood to get an estimate for m in terms of observed values $x_1, x_2, \ldots, x_n$ in a random sample.

4. Given n observed pairs $(x_1, y_1), (x_2, y_2), \ldots, (x_n, y_n)$, find a, b so that $\sum_{i=1}^{n} (y_i - a - bx_i)^2$ is minimum.

5. A coin is thrown 1000 times and a head arise 600 times. Use the principle of minimum chi-square to estimate the probability of a head.

6. Which is the better algorithm for finding a square root: the one based on the expansion for $(a + b)^2$ or the one based on Newton-Raphson method and why?

9.2 MATHEMATICAL MODELLING THROUGH CALCULUS OF VARIATIONS

9.2.1 Euler-Lagrange Equation

Consider
$$I = \int_{a}^{b} f\left(x, y, \frac{dy}{dx}\right) dx \tag{7}$$

For every well-behaved function y of x, we can find I as a real number so that I depends on what function y is of x. The problem of calculus of variations is to find that function $y(x)$ for which I is maximum or minimum. The answer is given by the solution of Euler-Lagrange's equation

$$\frac{\partial f}{\partial y} - \frac{d}{dx}\left(\frac{\partial f}{\partial y'}\right) = 0, \tag{8}$$

which is an ordinary differential equation of the second order. A proof of this result will be obtained in the next section by using dynamic programming.

If
$$I = \iint f\left(x, y, z, \frac{\partial z}{\partial x}, \frac{\partial z}{\partial y}\right) dx \, dy \equiv \iint f(x, y, z, p, q) \, dx \, dy, \tag{9}$$

then I is maximum or minimum when

$$\frac{\partial f}{\partial z} - \frac{\partial}{\partial x}\left(\frac{\partial f}{\partial p}\right) - \frac{\partial}{\partial y}\left(\frac{\partial f}{\partial q}\right) = 0 \tag{10}$$

9.2.2 Maximum-Entropy Distributions

(a) We want to find that probability distribution for a variate varying over the range $(-\infty, \infty)$ which has maximum entropy out of all distributions having a given mean m and a given variance σ^2.

Let $f(x)$ be the probability density function, then we have to maximize entropy defined by

$$S = -\int_{-\infty}^{\infty} f(x) \ln f(x) \, dx \tag{11}$$

subject to
$$\int_{-\infty}^{\infty} f(x) \, dx = 1, \quad \int_{-\infty}^{\infty} x f(x) \, dx = m,$$

$$\int_{-\infty}^{\infty} x^2 f(x) \, dx = \sigma^2 + m^2 \tag{12}$$

We form the Lagrangian

$$L = \int_{-\infty}^{\infty} -f(x) \ln f(x) - \lambda \int_{-\infty}^{\infty} f(x) \, dx - \mu \int_{-\infty}^{\infty} xf(x) \, dx - \nu \int_{-\infty}^{\infty} x^2 f(x) \, dx$$

$$(10)$$

Here the integrand contains only x and $y(=f(x))$ and there is no y' in it. As such (8) gives

$$-(1 + \ln f(x)) - \lambda - \mu x - \nu x^2 = 0 \qquad (14)$$

or

$$f(x) = Ae^{\mu x + \nu x^2} \qquad (15)$$

We use (11) to calculate A, μ, ν to get

$$f(x) = \frac{1}{\sqrt{2\pi}\sigma} e^{-[1/2(x - m)^2/\sigma^2]} \qquad (16)$$

This shows that out of all distributions with a given mean m and a given variance σ^2, the normal distribution $N(m, \sigma^2)$ has the maximum entropy.

Now mean and variance are simplest moments and the maximum entropy distribution for which these moments have prescribed values is the normal distribution. This gives one reason for the importance of the normal distribution.

(b) We now want to find the distribution over the interval $[0, \infty)$ which has maximum-entropy, out of all those which have given arithmetic and geometric means.

Here we have to maximize

$$-\int_0^{\infty} f(x) \ln f(x) \, dx \qquad (17)$$

subject to

$$\int_0^{\infty} f(x) \, dx = 1, \quad \int_0^{\infty} xf(x) \, dx = m, \quad \int_0^{\infty} \ln x f(x) \, dx = \ln g \qquad (18)$$

Using Lagrange's method and (8) we get

$$f(x) = A e^{-ax} x^{\gamma - 1} \qquad (19)$$

A, a, γ are determined by using (8). Thus gamma distribution has the maximum entropy out of all distributions which have given arithmetic and geometric means.

(c) We want to find the maximum entropy bivariate distribution when x, y vary from $-\infty$ to ∞ and when means, variances and covariance are prescribed.

We have to maximize

$$-\int_{-\infty}^{\infty} \int_{-\infty}^{\infty} f(x, y) \ln f(x, y) \, dx \, dy \qquad (20)$$

subject to

$$\int_{-\infty}^{\infty}\int_{-\infty}^{\infty} f(x, y)\, dx\, dy = 1, \int_{-\infty}^{\infty}\int_{-\infty}^{\infty} xf(x, y)\, dx\, dy = m_1,$$

$$\left\|\int_{-\infty}^{\infty}\int_{-\infty}^{\infty} yf(x, y)\, dx\, dy = m_2, \int_{-\infty}^{\infty}\int_{-\infty}^{\infty} x^2 f(x, y)\, dx\, dy = \sigma_1^2 + m_1^2,\right.$$

$$\int_{-\infty}^{\infty}\int_{-\infty}^{\infty} y^2 f(x, y)\, dx\, dy = \sigma_2^2 + m_2^2 \qquad (21)$$

$$\int_{-\infty}^{\infty}\int_{-\infty}^{\infty} xy f(x, y)\, dx\, dy = \rho\sigma_1\sigma_2 + m_1 m_2$$

Forming the Lagrangian and using (10), we get

$$f(x, y) = A\, e^{-a_1 x - a_2 y - b_1 x^2 - b_2 y^2 - cxy} \qquad (22)$$

Using (21) to find a_1, a_2, b_1, b_2, c, we get

$$f(x, y) = \frac{1}{2\pi\sigma_1\sigma_2\sqrt{1 - \rho^2}} \exp\left(-\frac{1}{2(1 - \rho^2)}\left(\frac{(x - m_1)^2}{\sigma_1^2}\right.\right.$$

$$\left.\left.-2\rho\frac{(x - m_1)(y - m_2)}{\sigma_1\sigma_2} + \frac{(y - m_2)^2}{\sigma_2^2}\right)\right)$$

which gives the density function for the bivariate normal distribution, so that out of all bivariate probability distributions for which x, y vary from $-\infty$ to ∞ and which have given means, variances and covariance, the distribution with the maximum entropy is the bivariate normal distribution.

(d) We want to find the multivariate distribution for $x_1, x_2, \ldots, x_n$ where

$$0 \leqslant x_1 \leqslant 1, 0 \leqslant x_2 \leqslant 1, \ldots, 0 \leqslant x_n \leqslant 1; x_1 + x_2 + \ldots + x_n = 1 \qquad (24)$$

for which $E(\ln x_1), \ldots, E(\ln x_n)$ have prescribed values and for which entropy is maximum.

Using the principle of maximum entropy, we get,

$$f(x_1, x_2, \ldots, x_n) = \frac{T(m_1 + m_2 + \ldots + m_n)}{T(m_1)T(m_2)\ldots, T(m_n)} x_1^{m_1-1} x_2^{m_2-1} \ldots x_{n-1}^{m_{n-1}-1}$$

$$(1 - x_1 - x_2 \ldots - x_{n-1})^{m_n-1} \qquad (25)$$

which is Dirichlet distribution.

9.2.3 Mathematical Modelling of Geometrical Problems through Calculus of Variations

(a) *Finding the path of shortest distance between two points in a plane*

Here $\qquad I = \int_a^b \sqrt{1 + \left(\frac{dy}{dx}\right)^2}\, dx, f(x, y, y') = \sqrt{1 + y'^2} \qquad (26)$

(8) gives $\qquad \frac{d}{dx}(y') = 0, \quad y' = \text{const}, y = mx + c \qquad (27)$

Alternatively $I = \int_{\theta_1}^{\theta_2} \sqrt{r^2 + \left(\dfrac{dr}{d\theta}\right)^2} \, d\theta, \quad f(\theta, r, r') = \sqrt{r^2 + r'^2}$ (28)

(8) gives $r\dfrac{d\theta}{dr} = \text{const.}, \quad \tan\varphi = \text{const.}, \quad \varphi = \text{const.}$ (29)

Thus the path of shortest distance between two points is a straight line.

(b) *Finding geodesics (paths of shortest distance) between two given points on the surface of a sphere*

Let $x = a \sin\theta \cos\varphi, \; y = a \sin\theta \sin\varphi, \; z = a\cos\theta,$ (30)

then $I = \int_{x_0, y_0, z_0}^{x_1, y_1, z_1} \sqrt{(dx)^2 + (dy)^2 + (dz)^2}$

$$= a\int_{\varphi_1}^{\varphi_2} \sqrt{\sin^2\theta + \left(\dfrac{d\theta}{d\varphi}\right)^2} \, d\varphi$$ (31)

$$f = \sqrt{\sin^2\theta + \theta'^2}$$ (32)

(8) gives $\dfrac{d\varphi}{d\theta} = \dfrac{\sin\alpha}{\sin\theta\sqrt{\sin^2\theta - \sin^2\alpha}}$ (33)

Integrating $\tan\alpha\cos\theta - \sin\theta\cos\varphi\cos\beta + \sin\theta\sin\varphi\sin\beta = 0$ (34)

or $z\tan\alpha - x\cos\beta + y\sin\beta = 0,$ (35)

which is the equation of a plane passing through the centre of the sphere. Hence a geodesic is a great circle arc passing through the two given points.

(c) *Finding Minimal surface of revolution* i.e. finding the equation of a curve joining two given points in a plane, which when rotated about the x-axis gives a surface with minimum area.

The surface area is given by

$$S = 2\pi \int_a^b y \, ds$$

$$= 2\pi \int_a^b y \sqrt{1 + \left(\dfrac{dy}{dx}\right)^2} \, dx$$ (36)

$$f(x, y, y') = y\sqrt{1 + y'^2}$$ (37)

(8) gives $y\sqrt{1 + y'^2} = \text{constant}$ (38)

Integrating $y = c \cosh(x/c)$ (39)

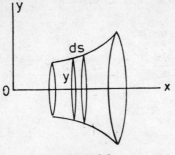

Figure 9.2

Thus the minimal surface of revolution is the catenoid obtained by rotating a catenary about its directrix. A related problem was solved in section 6.4.

The soap film between two loops of circular wire is a practical example of a catenoid. As we go on increasing the distance between the loops, a stage comes when the film breaks down. This corresponds to the case when no catenoid is possible.

Figure 9.3

(d) *Determining a given plane closed curve with given perimeter enclosing maximum area* (The isoperimetric curve)

Using polar coordinates, we have to maximize

$$I = \frac{1}{2} \int_0^{2\pi} r^2 \, d\theta \tag{40}$$

subject to

$$\int_0^{2\pi} \sqrt{r^2 + \left(\frac{dr}{d\theta}\right)^2} \, d\theta = \text{constant} \tag{41}$$

Using Lagrange's method,

$$f = \frac{1}{2} r^2 - \lambda \sqrt{r^2 + r'^2} \tag{42}$$

(8) gives

$$\frac{1}{2} r^2 - \frac{\lambda r^2}{\sqrt{r^2 + r'^2}} = \text{constant} \tag{43}$$

Differentiation with respect to θ, gives

$$\frac{r^2 + r'^2 - rr''}{(r^2 + r'^2)^{3/2}} = \frac{1}{\lambda}, \tag{44}$$

but the LHS is the expression for the curvature of the curve. As such the required curve is a curve of constant curvature i.e. it is a circle.

The problem is supposed to have arisen from the gift of a king who was happy with a person and promised to give him all the land he could enclose by running round in a day. Since he could run a fixed distance, the perimeter of his path was fixed and as such the radius of the circle he should describe is known.

(e) *Finding the solid of revolution with given surface area and maximum volume*

If V is the volume and S is the surface area

$$V = \pi \int y^2 \, dx, \quad S = 2\pi \int y \sqrt{1 + \left(\frac{dy}{dx}\right)^2} \, dx \tag{45}$$

$$f(x, y, y') = \pi y^2 - 2\lambda \pi y \sqrt{1 + y'^2} \tag{46}$$

(8) gives,

$$y^2 - \frac{2\lambda y}{\sqrt{1 + y'^2}} = \text{constant} \tag{47}$$

Its integration for general values of the constant involves elliptic functions, but for the special case when the constant is taken as zero, (47) gives

$$y = 2\lambda \cos \psi \quad \text{so that} \quad \sin \psi = \frac{dy}{ds} = -2\lambda \sin \psi \frac{d\psi}{ds} \tag{48}$$

or $\quad \dfrac{d\psi}{ds} = -\dfrac{1}{2\lambda} = \text{constant},$ \hfill (49)

so that in this case the surface is obtained by rotating a circle and is thus a sphere.

9.2.4 Mathematical Modelling of Situations in Mechanics Through Calculus of Variations

(a) *Finding the shape of a freely hanging uniform heavy string under gravity when the two ends of it are fixed*

We minimize the potential energy V subject to length of the string being fixed. As such we have to minimize

$$V = mg \int y \sqrt{1 + y'^2} \, dx \tag{50}$$

subject to $\qquad l = \int \sqrt{1 + y'^2} \, dx$ \hfill (51)

Therefore $\qquad f = y \sqrt{1 + y'^2} - \lambda \sqrt{1 + y'^2}$ \hfill (52)

(8) gives $\qquad \dfrac{dy}{dx} = \left(\dfrac{(y - \lambda)^2}{c^2} - 1 \right)^{\frac{1}{2}}$ \hfill (53)

Integrating $\qquad y - \lambda = \cosh \dfrac{x - a}{c},$ \hfill (54)

so that the required curve is a catenary

(b) *Finding the equation of the smooth vertical curve along which the time of descent under gravity between any two given points is minimum* (Brachistrochrone Problem)

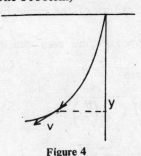

Figure 4

Using the principle of conservation of energy, we get

$$\frac{1}{2} mv^2 - mgy = \text{constant}$$

If the particle starts from rest when $y = 0$ (Fig. 9.4), we get

$$v^2 = 2gy \quad \text{or} \quad \frac{ds}{dt} = \sqrt{2gy} \tag{55}$$

or $\qquad T = \int \dfrac{ds}{\sqrt{2gy}} = \dfrac{1}{\sqrt{2g}} \int_a^b \dfrac{1}{\sqrt{y}} \sqrt{1 + \left(\dfrac{dy}{dx} \right)^2} \, dx$ \hfill (56) so that $\qquad f(x, y, y') = \sqrt{1 + y'^2}/\sqrt{y}$ $\qquad\qquad$ (57)

(8) gives $\qquad y(1 + y'^2) = 2c$ $\qquad\qquad$ (58)

or $\qquad\qquad y = c(1 + \cos 2\psi)$ $\qquad\qquad$ (59)

Now $\qquad dx = \cot \psi \, dy = -4c \cos^2 \psi \, d\psi$ $\qquad\qquad$ (60)

$\qquad\qquad x = a - c(2\psi + \sin 2\psi)$ $\qquad\qquad$ (61)

Equations (59) and (61) give the parameteric equations of a cycloid.

(c) *Discussion of the shapes of vibrating strings and membranes*

These have already been discussed in section 6.4.

(d) *Obtaining the equation of the free surface of a fluid rotating in a cylinder about its axis under gravity*

Consider the element of volume $2\pi x \, dx \, dz$ and mass $\rho 2\pi x \, dx \, dz$. Its potential energy is

$$\rho 2\pi x \, dx \, dz \left(gz - \frac{1}{2}\omega^2 x^2 + c \right),$$
$$\qquad\qquad\qquad (62)$$

so that the total potential energy of the fluid is

$$2\pi\rho \int_{x=0}^{a} \int_{z=0}^{y} x$$

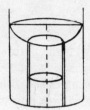

Figure 5

$$\left(gz - \frac{1}{2}\omega^2 x^2 + c \right) dx \, dz$$

$$= \pi\rho \int_{0}^{a} (gy^2 - \omega^2 x^2 y + 2cy)x \, dx \qquad (63)$$

Since potential energy has to be minimum, we minimize (63). Here

$$f = (gy^2 - \omega^2 x^2 y + 2cy)x \qquad\qquad (64)$$

(8) gives $\qquad\qquad 2gy - \omega^2 x^2 + 2c = 0,$ $\qquad\qquad$ (65)

which is a parabola. so that the free surface is a paraboloid of revolution

(e) *Lagrange's equations of Motion*

Let $q_1, q_2, \ldots, q_n$ be 'generalised' coordinates in terms of which a dynamical system is described, then its kinetic energy T is a function of q_1, q_2, $\ldots, q_n$; $q'_1, q'_2, \ldots, q'_n$ and its potential energy V is a function of $q_1, q_2, \ldots, q_n$ only. According to Hamiltonian principle, we then have to find an extreme value for

$$H = \int (T(q_1, q_2, \ldots, q_n; q'_1, \ldots, q'_n) - V(q_1 q_2, \ldots, q_n)) \, dt \qquad (66)$$

Using equation similar to (8) for $q_1, q_2, \ldots, q_n$, we get

$$\frac{\partial T}{\partial q_i} - \frac{\partial V}{\partial q_i} - \frac{d}{dt}\left(\frac{\partial T}{\partial q'_i}\right) = 0, \quad i = 1, 2, \ldots n \qquad (67)$$

or $\qquad\qquad \dfrac{d}{dt}\left(\dfrac{\partial T}{\partial q'_i}\right) - \dfrac{\partial T}{\partial q_i} = -\dfrac{\partial V}{\partial q_i}, \quad i = 1, 2, \ldots, n \qquad (68)$

or
$$\frac{d}{di}\left(\frac{\partial L}{\partial q_2'}\right) - \frac{\partial L}{\partial q_i} = 0; \ L = T - V; \ i = 1, 2, \ldots, n \qquad (69)$$

Equations (67) or (68) or (69) are called Lagrange's equations of motion. These are n simultaneous ordinary differential equations of second order for determining $q_1, q_2, \ldots, q_n$ as functions of t.

9.2.5 Mathematical Modelling in Bioeconomics Through Calculus of Variations

Mathematical Bioeconomics is an interdisciplinary subject in which we use mathematical methods to optimize the economic profits from the utilization of renewable biological resources like forests and fisheries.

Let $x(t)$ be the fish population at time t and let $h(t)$ be the rate at which it is harvested, then we get the equation

$$\frac{dx}{dt} = F(x) - h(t), \qquad (70)$$

where $F(x)$ is the natural biological rate of growth. Let $c(x)$ be the cost of harvesting a unit of fish when the population size is $x(t)$ and let p be the selling price per unit fish so that the profit per unit fish is $(p - c(x))$ and the profit in time interval $(t, t + dt)$ is $(p - c(x)\ h(t)\ dt$. If δ is the instantaneous discount rate, the present value of the total profit is

$$P = \int_0^\infty e^{-\delta t}(p - c(x))h(t)\ dt \qquad (71)$$

If we know $h(t)$, we can use (70) to solve for $x(t)$ and then we can use (71) to determine P so that P depends on what function h is of t. We have to determine that function $h(t)$ for which P is maximum. Substituting for $h(t)$ from (70) in (71), we get

$$P = \int_0^\infty e^{-\delta t}(p - c(x))(F(x) - x')\ dt \qquad (72)$$

so that
$$f(t, x, x') = e^{-\delta t}(p - c(x))(F(x) - x') \qquad (73)$$

Using Euler-Lagrange equation (8),

$$\frac{\partial f}{\partial x} - \frac{d}{dt}\left(\frac{\partial f}{\partial x'}\right) = 0$$

or
$$e^{-\delta t}(-c'(x))(F(x) - x') + e^{-\delta t}(p - c(x))(F'(x)$$
$$-\frac{d}{dt}\left[e^{-\beta t}(c(x) - p)\right] = 0 \qquad (74)$$

or
$$-c'(x)(F(x) - x') + (p - c(x))F'(x) + \delta(c(x) - p)$$
$$- c'(x)x' = 0$$

or
$$-c'(x)F(x) + (p - c(x))(F'(x) - \delta) = 0 \qquad (75)$$

which determines a constant value x^* for x and then (70) gives the rate of harvesting as constant and equal to $F(x^*)$.

If the initial population is less than x^*, we should do no harvesting till the population rises to x^* and then begin harvesting at a constant rate $F(x^*)$. If the initial population is more than x^*, we should do harvesting at the maximum permissible rate till the population falls to x^*, and then begin doing harvesting at a constant rate $F(x^*)$.

9.2.6 Mathematical Modelling in Optics Through Calculus of Variations

According to Fermat's principle of least time, light travel from a given point A to another point B in such a way as to take the least possible time. If $\mu(x, y)$ is the refractive index at the point (x, y), then the velocity of light at the point is $c/\mu(x, y)$ and the time taken in going from A to B is

$$= \int_A^B \frac{ds}{c/\mu} = \int_A^B \mu(x, y) \sqrt{1 + \left(\frac{dy}{dx}\right)^2}\, dx \qquad (76)$$

$\therefore \qquad f(x, y, y') = \mu(x, y)\sqrt{1 + y'^2} \qquad (77)$

(8) gives

$$\frac{\partial \mu}{\partial y}\sqrt{1 + y'^2} - \frac{d}{dx}\left[\mu\, \frac{y'}{\sqrt{1 + y'^2}}\right] = 0 \qquad (78)$$

or

$$\frac{\partial \mu}{\partial y} = \frac{d}{ds}\, (\mu \sin \psi) \qquad (79)$$

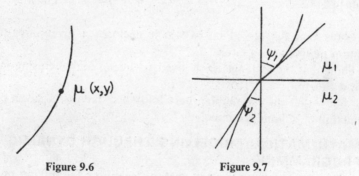

Figure 9.6   Figure 9.7

If y-axis separates two media of refractive indices μ_1 and μ_2, then

$$\frac{\partial \mu}{\partial y} = 0$$

and so

$$\mu_1 \sin \psi_1 = \mu_2 \sin \psi_2 \qquad (80)$$

which is Snell's law of refraction

EXERCISE 9.2

1. Prove that (8) is an ordinary differential equation and (10) is a partial differential equation and both are of second order.

2. Find the maximum-entropy distributions when

(i) range is $[0, 1]$ and $E(\ln x)$, $E(\ln (1 - x))$ are prescribed.

(ii) range is $[0, \infty)$ and $E(x)$ is prescribed

(iii) range is $[0, \infty)$ and $E(\ln x)$ and $E(\ln (1 + x))$ are prescribed

(iv) range is $[0, \infty)$ and $E(\ln x)$ and $E(\ln x)^2$ are prescribed.

3(a) Find A, μ, ν in (15) in terms of m, σ^2.

(b) Find A, a_1, a_2, b_1, b_2, c in (22) in terms of $m_1, m_2, \sigma_1^2, \sigma_2^2$ and ρ.

(c) Prove (25).

4. Find the equilibrium shape of uniform heavy string overhanging two smooth pulleys by minimizing the potential energy of the string.

5. For the condition satisfied for the extremum of

$$I = \int h(x, y)(1 + y'^2)^{1/2} \, dx$$

Discuss the special case when

$$h(x, y) = y^r \quad \text{and} \quad r = 1, -1, \frac{1}{2}, -\frac{1}{2}.$$

6. Find u such that the average value of $(\Delta u)^2$ over a certain region is constant.

7. Maximize
$$I = \frac{1}{2} \int_{t_1}^{t_2} (xy' - yx') \, dt$$

subject to
$$J = \int_{t_1}^{t_2} (x'^2 + y'^2)^{1/2} \, dt = \text{const.}$$

8. Show that the closed curve which encloses a given area and has minimum perimeter is a circle.

9. Show that the rectangle with given perimeter and enclosing maximum area is a square.

10. Show that the rectangular parallelopiped having a given perimeter and maximum volume is a cube.

9.3 MATHEMATICAL MODELLING THROUGH DYNAMIC PROGRAMMING

Dynamic programming is an important technique for solving multi-stage optimization mathematical modelling problems. The main principle used is the principle of optimality discussed in sections 6.4 and 9.1.

Quite often the problem of maximising of a function of n variables can be reduced to an n-stage decision problem where a stage corresponds to the choice of the optimizing values of a variable. Instead of dealing with one problem of maximizing a function of n variables, we deal with n problems of maximizing a function of one variable and we deal with these in a sequence. This leads to a considerable simplification of the problem, as we shall see in the examples below:

9.3.1 Two Classes of Optimization

(a) A Class of Maximization Problem

We have to allocate a total resource c to n activities so as to maximize the total output when the output from the ith activity when an amount x_i

is allotted to it is $g_i(x_i)$ where $g_i(x_i)$ is a concave function of x_i so that our problem is

maximize $\qquad\qquad g_1(x_1) + g_2(x_2) + \ldots + g_n(x_n)$ $\qquad\qquad$ (81)

subject to $\qquad\qquad x_1 + x_2 + \ldots + x_n = c;$

$$x_i \geqslant 0, \quad i = 1, 2, \ldots n \qquad\qquad (82)$$

Let $f_n(c)$ be the maximum value, then the principle of optimality gives

$$f_n(c) = \max_{0 \leqslant x_n \leqslant c} (g_n(x_n) + f_{n-1}(c - x_n)) \qquad\qquad (83)$$

Also $\qquad\qquad f_1(c) = g_1(c) \qquad\qquad$ (84)

so that $\qquad\qquad f_2(c) = \max_{0 \leqslant x_2 \leqslant c} (g_2(x_2) + g_1(c - x_2)) \qquad\qquad$ (85)

The function to be maximized is the sum of two concave function and its maximum arises when

$$g_2'(x_2) = g_1'(c - x_2) \qquad\qquad (86)$$

Thus x_2 is known and therefore $f_2(c)$ is determined for all values of c. In particular if $g_1(x) = g_2(x) = g(x)$, then $g(x_1) + g(x_2)$ is maximum when $x_1 = x_2 = c/2$ and the maximum value is $2g(c/2)$. Similarly if $g(x)$ is concave, then the maximum value of $g(x_1) + g(x_2) + \ldots + g(x_n)$ occurs when

$$x_1 = x_2 = \ldots = x_n = \frac{c}{n}$$

and the maximum value is $ng(c/n)$. For a general value of n, this result can be established by mathematical induction.

Special Cases

(i) Since $\ln x$ is a concave function,
$\ln x_1 + \ln x_2 + \ldots + \ln x_n$ is maximum subject to $x_1 + x_2 + \ldots + x_n = c$ when $x_1 = x_2 = \ldots x_n = c/n$ and the maximum value is $n \ln c/n$ and the maximum value of $x_1 x_2 \ldots x_n$ is $(c/n)^n$.

(ii) Since $-x \ln x$ is a concave function,
$-\left(\sum\limits_{i=1}^{n} p_i \ln p_i\right)$ is maximum subject to $\sum\limits_{i=1}^{n} p_i = 1$ when $p_1 = p_2 = \ldots = p_n = 1/n$.

(iii) Since $(x^\alpha - x)/(1 - \alpha)$ is concave function
$(\sum\limits_{i=1}^{n} p_i^\alpha - 1)/(1 - \alpha)$ is maximum subject to $\sum\limits_{i=1}^{n} p_i = 1$ when $p_1 = p_2 = \ldots = P_n = 1/n$.

(iv) Since $-x \ln x + \dfrac{1}{a}(1 + ax) \ln (1 + ax)$ is a concave function,
$-\sum\limits_{i=1}^{n} p_i \ln p_i + \dfrac{1}{a} \sum\limits_{i=1}^{n} (1 + ap_i) \ln (1 + ap_i)$ is maximum subject to $\sum\limits_{i=1}^{n} p_i = 1$, when $p_1 = p_2 = \ldots = p_n = 1/n$.

All these results are of considerable importance in applications of information theory.

(b) A Class of Minimization Problems

If $h(x)$ is a convex function, the functional equation for obtaining the minimum value of $h(x_1) + h(x_2) + \ldots + h(x_n)$ subject to

$$x_1 + x_2 + \ldots + x_n = c, \quad x_1 \geqslant 0, \quad x_2 \geqslant 0, \ldots, x_n \geqslant 0$$

is given by

$$f_n(c) = \min_{0 \leqslant x_n \leqslant c} (h(x_n) + f_{n-1}(c - x_n)) \tag{87}$$

and proceeding as before we find that the minimum value of

$$h(x_1) + h(x_2) + \ldots + h(x_n) \tag{88}$$

subject to (82) is $nh(c/n)$ and occurs when $c_1 = c_2 = \ldots = c_n = 1/n$

In the same way if $h_1(x_1), h_2(x_2), \ldots, h_n(x)$ are all convex functions, the minimum value of $h(x_1) + h_2(x_2) + \ldots + h_n(x_n)$ subject to (82) occurs when

$$h_1'(x_1) = h_2'(x_2) = \ldots = h_n'(x_n). \tag{89}$$

Special Cases

(i) Since $x \ln \dfrac{x}{y}$ is a convex function of x, the minimum value of

$\sum\limits_{i=1}^{n} x_i \ln \dfrac{x_i}{y_i}$ subject to $\sum\limits_{i=1}^{n} x_i = c$, $\sum\limits_{i=1}^{n} y_i = d$ occurs when

$$1 + \ln \frac{x_1}{y_1} = 1 + \ln \frac{x_2}{y_2} = \ldots = 1 + \ln \frac{x_n}{y_n} \tag{90}$$

or

$$\frac{x_1}{y_1} = \frac{x_2}{y_2} = \ldots = \frac{x_n}{y_n} = \frac{c}{d} \tag{91}$$

and the minimum value of $\sum\limits_{i=1}^{n} x_i \ln \dfrac{x_i}{y_i}$ is $c \ln \dfrac{c}{d}$. If $\sum\limits_{i=1}^{n} x_i = \sum\limits_{i=1}^{n} y_i$ then the minimum value is zero.

(ii) Since $(x^\alpha y^{1-\alpha} - y)/(\alpha - 1)$ is a convex function of x, the quantity $\sum\limits_{i=1}^{n} (x_i^\alpha y_i^{1-\alpha} -)/(\alpha - 1)$ is minimum when (90) is satisfied and its minimum value is $((c/d)^\alpha c - d)/(\alpha - 1)$ and if $c = d$, the minimum value is zero.

9.3.2 Some Other Allocation Problems

(a) A Cargo-Loading Problem

We consider a vessel whose maximum cargo capacity is Z tons. Let v_i and w_i denote respectively the value end weight of the ith item and let x_i denote the number of items of type i chosen. The problem of determining the most valuable cargo consists in maximizing

$$L_n(X) = \sum_{i=1}^{n} x_i v_i \tag{92}$$

subject to

$$\sum_{i=1}^{n} x_i w_i \leqslant Z, \quad x_i = 0, 1, 2, \ldots \tag{93}$$

Let $f_n(Z)$ denote the maximum value, then

$$f_1(Z) = \left[\frac{Z}{w_1}\right] v_1, \tag{94}$$

where $[y]$ denotes the greatest integer less than or equal to y. The principle of optimality then gives

$$f_n(Z) = \max_{x_n} [x_n v_n + f_{n-1}(Z - x_n w_n)] \tag{95}$$

where the maximization with respect to x_n is over the set of values

$$x_n = 0, 1, 2, \ldots \left[\frac{Z}{w_n}\right] \tag{96}$$

This is essentially a problem of linear integer programming which we have solved by using dynamic programming technique.

(b) Reliability of Multicomponent Devices

We consider an equipment containing n components in series so that if one component fails, the whole equipment fails. For ensuring greater reliability of the equipment, we provide duplicate components in parallel at each stage. We assume that the units in each stage are supplied with switching circuits which have the property of shunting a new component into the circuit when an old one fails. We want to choose the number of components at each stage so that the probability of successful operation of the system is maximum subject to a given amount of money being available for duplicate components.

Let $\varphi_j(m_j)$ denote the probability of successful operation of the system when m_j components are used at the jth stage. Let c_j be the cost of a single component at the jth stage so that we have the constraint

$$\sum_{j=1}^{n} m_j c_j \leqslant c \tag{97}$$

The reliability of the n-stage equipment i.e. the probability of its successful operation is given by

$$\prod_{j=1}^{n} \varphi_j(m_j). \tag{98}$$

Let its maximum value, which depends on c and n be denoted by $f_n(c)$, then by the principle of optimality

$$f_n(c) = \max_{m_n} [\varphi_n(m_n) f_{n-1}(c - c_n m_n)], \tag{99}$$

where m_n can take value $0, 1, 2, \ldots [c/c_n]$. Also

$$f_1(c) = \varphi_1\left(\left[\frac{c}{m_1}\right]\right) \tag{100}$$

(c) A Farmer's Problem

A farmer starts with q tons of wheat. He can sell a part, say y tons for an amount $g(y)$ and he can sow the remaining $q - y$ tons and get a $(q - y)$ tons $(a \geqslant 1)$ out of it for further selling and sowing. It is required to find the optimum policy for him if he intends to remain in business for n years. Let $f_n(q)$ be the maximum return on following an optimum policy, then by the principle of optimality

$$f_n(q) = \max_{0 \leqslant y \leqslant q} (g(y) + f_{n-1}(a(q - y))) \tag{101}$$

and

$$f_1(q) = \max_{0 \leqslant y \leqslant q} (g(y) + g(a(q - y))) \tag{102}$$

For an infinite stage process, applying the limiting process to (101), we get

$$f(q) = \max_{0 \leqslant y \leqslant q} (g(y) + f(a(q - y))) \tag{103}$$

which is a functional equation to solve for $f(q)$.

(d) A purchase problem

An amount x can be used to buy two equipments A and B. If an amount y is invested in type A, we get $g(y)$ hours of useful work in the course of a year and the equipment has a salvage value ay $(0 < a < 1)$. The remaining amount $x-y$ invested in equipment of type B gives $h(x - y)$ hours of useful work and has a salvage value $b(x - y)$ $(0 < b < 1)$. If $f_n(x)$ is the number of useful hours on following an optimal policy, we get

$$f_n(x) = \max_{0 \leqslant y \leqslant x} (g(y) + h(x - y) + f_{n-1}(ay + bx - by)), \tag{104}$$

$$f_1(x) = \max_{0 \leqslant y \leqslant x} (g(y) + h(x - y)) \tag{105}$$

If the infinite-period optimal policy gives $f(x)$ as the number of useful hours of work, then taking the limit of (104) we get

$$f(x) = \max_{0 \leqslant y \leqslant x} (g(y) + h(x - y)) + f(ay - bx - by)) \tag{106}$$

(e) Allocation Processes involving two types of resources

Suppose we have two types of resources in quantities x and y respectively. We have to allocate these resources to n activities and if we allocate, x_i, y_i to ith activity, the return is given by $g_i(x_i, y_i)$ so that the total return is

$$\sum_{i=1}^{n} g_i(x_i, y_i) \tag{107}$$

Let $f_n(x, y)$ be the maximum return for n activities following an optimal policy, then the principle of optimality gives

$$f_n(x, y) = \max_{0 \leqslant x_n \leqslant x} \max_{0 \leqslant y_n \leqslant y} (g_n(x_n, y_n) + f_{n-1}(x_n - x_n, y - y_n)), \; n \geqslant 2 \tag{108}$$

$$f_1(x, y) = g_1(x, y). \tag{109}$$

(f) *Transportation Problem*

We have m origins $0_1, 0_2, \ldots, 0_m$ where quantities $x_1, x_2, \ldots, x_n$ of a certain commodity are available and these have to be supplied to n destinations $D_1, D_2, \ldots D_n$ where quantities $y_1, y_2, \ldots y_n$ are required. We further assume that

$$\sum_{i=1}^{m} x_i = \sum_{j=1}^{n} y_j \qquad (110)$$

The cost of transporting x_{ij} commodities from ith origin to jth destination is $g_{ij}(x_{ij})$ so that we have to minimize

$$\sum_{j=1}^{n} \sum_{i=1}^{m} g_{ij}(x_{ij}) \qquad (111)$$

subject to

$$x_{ij} \geqslant 0, \ \sum_{j=1}^{n} x_{ij} = x_i, \ \sum_{i=1}^{m} x_{ij} = y_j, \ \sum_{i=1}^{m} x_i = \sum_{j=1}^{n} y_j \qquad (112)$$

Let $f_n(x_1, x_2, \ldots, x_m)$ denote the minimal cost obtained by following an optimal policy, then the principle of optimality gives

$$f_n(x_1, x_2, \ldots, x_m) = \min_{R_n} ((g_{1n}(x_{1n}) + g_{2n}(x_{2n}) + \ldots + g_{mn}(x_{nm})$$

$$+ f_{n-1}(x_1 - x_{1n}, x_2 - x_{2n}, \ldots, x_m - x_{mn})), \qquad (113)$$

where R_n is the m-dimensional region determined by

$$0 \leqslant x_{in} \leqslant x_i, (i = 1, 2, \ldots, m), \ \sum_{i-1}^{m} x_{in} = y_n \qquad (114)$$

Instead of dealing with mn independent variables x_{ij} at one time, we have to minimize with respect to variations in m variables at a time and the reduction in dimensionality is quite significant. Yet for $m > 2$, the problem of computation is still difficult. For $m = 2$ i.e. for the case of two origins, we get

$$f_n(x_1, x_2) = \min_{0 \leqslant x_{1n} \leqslant x_1} (g_{1n}(x_{1n}) + g_{2n}(y_2 - x_{1n})$$

$$+ f_{n-1}(x_1 - x_{1n}, x_2 - y_2 + x_{1n})) \qquad (115)$$

which is more easily solvable.

9.3.3 Dynamic Programming and Calculus of Variations

Let

$$I = \int_{x,y}^{x_0, y_0} F\left(x, y, \frac{dy}{dx}\right) dx, \qquad (116)$$

then the value of I depends on what function y is of x, the starting point x, y and the final point x_0, y_0. If we choose different functions $y(x)$ and find the minimum value of I, this minimum value will depend on x, y and x_0, y_0. If we keep x_0, y_0 fixed, the minimum value will depend on x, y only. Let $f(x, y)$ be this minimum value.

To apply dynamic programming, we break up the interval (x, x_0) into two parts $(x, x + \Delta x)$ and $(x + \Delta x, x_0)$. In the first interval, we choose an arbitrary slope y', so that the contribution of the first interval to I is

$$\int_x^{x+\Delta x} F(x, y, y')\, dx = F(x, y, y')\Delta x + 0(\Delta x)^2 \tag{117}$$

The starting point for the second interval is $x + \Delta x$, $y + y'\Delta x$ and for this interval, we use the optimal policy to get

$$f(x + \Delta x, y + y'\Delta y) = f(x, y) + \Delta x \frac{\partial f}{\partial x} + y'\Delta x \frac{\partial f}{\partial y} + 0(\Delta x)^2 \tag{118}$$

Applying the principle of optimality, we get

$$f(x, y) = \min_{y'}\left[\Delta x F(x, y, y') + f(x, y) + \Delta x \frac{\partial f}{\partial x} \right.$$
$$\left. + y'\Delta x \frac{\partial f}{\partial y} + 0(\Delta x)^2\right] \tag{119}$$

Taking the limit as $\Delta x \to 0$

$$0 = \min_{y'}\left[F(x, y, y') + \frac{\partial f}{\partial x} + y'\frac{\partial f}{\partial y}\right] \tag{120}$$

For the expression within brackets to be minimum

$$0 = \frac{\partial F}{\partial y'} + \frac{\partial f}{\partial y} \tag{121}$$

When we solve for y' from (121) and substitute in (120) we get the minimum value of the expression as zero so that

$$0 = F(x, y, y') + \frac{\partial f}{\partial x} + y'\frac{\partial f}{\partial y} \tag{122}$$

From (121) and (122), we can determine

(i) y as a function of x and
(ii) $f(x, y)$ as a function of x, y.

Differentiating (121) totally with respect to x, we get

$$\frac{d}{dx}\left(\frac{\partial F}{\partial y'}\right) + \frac{\partial^2 f}{\partial x\, \partial y} + \frac{\partial^2 f}{\partial y^2}y' = 0 \tag{123}$$

Differentiating (122) partially with respect to y, we get

$$F_y + F_{y'}\frac{\partial y'}{\partial y} + \frac{\partial^2 f}{\partial x\, \partial y} + \frac{\partial^2 f}{\partial y^2}y' + \frac{\partial f}{\partial y}\frac{\partial y'}{\partial y} = 0 \tag{124}$$

Eliminating $f(x, y)$, we get Euler-Lagrange equation

$$\frac{d}{dx}\left(\frac{\partial F}{\partial y'}\right) - \frac{\partial F}{\partial y} = 0 \tag{125}$$

For the more general case when there are several dependent variables y_1, y_2, ..., y_n i.e. where we have to minimize

$$I = \int F(x, y_1, y_2, \ldots, y_n, y_1', y_2', \ldots, y_n') \, dx, \tag{126}$$

the equation corresponding to (120) is

$$0 = \min_{y_1', y_2', \ldots y_n'} \left[F + \frac{\partial f}{\partial x} + \sum_{j=1}^{n} y_j' \frac{\partial f}{\partial y_j} \right] \tag{127}$$

which gives the following two equations

$$\frac{\partial F}{\partial y_i'} + \frac{\partial f}{\partial y_i} = 0, \, i = 1, 2, \ldots, n \tag{128}$$

$$F + \frac{\partial f}{\partial x} + \sum_{j=1}^{n} y_j' \frac{\partial f}{\partial y_j} = 0 \tag{129}$$

Eliminating f, we get Euler's equations

$$\frac{d}{dx}\left(\frac{\partial F}{\partial y_i'}\right) - \frac{\partial F}{\partial y_i} = 0, \quad i = 1, 2, \ldots, n \tag{130}$$

9.3.4 Some Other Applications of Dynamic Programming

(a) *A Defective Coin Search Problem*

We consider the problem of using an equal arms balance to detect the only heavy coin in a lot of N coins of similar appearance. Let f_N denote the maximum number of weightings required using an optimal policy.

As each stage, we weigh one batch of k coins against another and observe the result. Either the two sets of coins will balance or they will not. If the two sets balance, the heavy coin must be in the remaining $N - 2k$ coins. If they do not balance, then we have already found the group of k coins to which it belongs. Thus

$$f_N = 1 + \min_{0 \leqslant k \leqslant N/2} \max [f_k, f_{N-2k}] \tag{131}$$

To minimize, we want k and $N - 2k$ to be as near as possible. Accordingly we take $k \doteq [N/3]$ or $[N/3] + 1$ depending on whether N has the form $3m + 1$ or $3m + 2$.

(b) *An Inventory Problem*

At the beginning of each period, a businessman raises his stock to y. There is no time lag between his ordering and supplies being received. The cost of ordering an amount z is $h(z)$. During a period, the probability that the demand lies between s and $s + ds$ is $\varphi(s)ds$. If the demand exceeds stocks, there is a penalty cost $p(z)$ associated with the shortage z. The businessman starts with a stock x and wants to continue in business for n periods. It is required to find y so that his cost of ordering and stock-out is minimized.

In the first period, he has to spend $k(y - x)$ on ordering new stocks. If the demand lies between s and $s + ds$, the expected stockout cost is $\int_y^\infty p(s - y)\varphi(s)\, ds$ since the cost will be there if $s \geqslant y$. Thus if $f_n(x)$ denotes the minimum cost for n periods,

$$f_1(x) = \min\left[k(y - x) + \int_y^\infty p(s - y)\varphi(s)\, ds\right] \tag{132}$$

For writing the general recurrence relation, we note that at the end of the first period, the stock may be zero with probability $\int_y^\infty \varphi(s)\, ds$ or it may be $y - s$ if the demand has been for s commodites in this period ($s \leqslant y$). The principle of optimality then gives

$$f_n(x) = \min_{y \geqslant x}\left[k(y - x) + \int_y^\infty p(s - y)\varphi(s)\, ds + f_{n-1}(0)\int_y^\infty \varphi(s)\, ds\right.$$
$$\left. + \int_0^y f_{n-1}(y - s)\varphi(s)\, ds\right] \tag{133}$$

(c) Optimal Exploitation of a Fishery Containing Many Interacting Species

Let $x_i(t)$ be the population of the ith species at time t and let $h_i(t)$ be its rate of harvesting at time t so that

$$\frac{dx_i}{dt} = a_i x_i - h_i(t); \quad i = 1, 2, \ldots, n, \tag{134}$$

Let $\quad h_i(t) = \alpha_i + \sum_{j=1}^n \beta_{ij} x_j + \gamma_i E, \quad i = 1, 2, \ldots, n \tag{135}$

where $E(t)$ is the effort per unit time. Let the cost of making as effort E be $bE^2 - kE - m$, then the present value of the profit is

$$P = \int_0^\infty e^{-\delta t}\{\sum_{i=1}^n p_i(\alpha_i + \sum_{j=1}^n \beta_{ij} x_j + \gamma_i E) - bE^2 - kE - m)\, dt, \tag{136}$$

where p_i is the selling price per unit of the ith species.

The maximum value of P depends on the initial population sizes of the species. Let this maximum value be $f(R_1, R_2, \ldots, R_n)$ where

$$X_i(0) = R_i, (i = 1, 2, \ldots, n) \tag{137}$$

We now split the integral in (136) into two, over the ranges 0 to Δ and Δ to ∞, where Δ is small. We choose some arbitrary value for the initial effort E and find the value of the first integral for this value of E because Δ is small. From (136), if the maximum value is $f(R_1, R_2, \ldots, R_n)$, then for the second integral, the maximum value is $f(R_1', R_2', \ldots, R_n')$ when $R_1', R_2', \ldots, R_n'$ are the population sizes at time Δ determined from (134), so that

$$R_i' = R_i + \Delta(a_i R_i - \alpha_i - \sum_{j=1}^n \beta_{ij} R_j - \gamma_i E) \tag{138}$$

We then find the sum of the first integral and the maximum value of the second integral. Both these depend on the choice of E. We now choose E so as to maximize the sum. This gives the equation

$$f(R_1, R_2, \ldots, R_n) = \max_E [\Delta\{\sum_{i=1}^{n} p_i(\alpha_i + \beta_{ij}R_j + \gamma_i E) - bE^2 - kE - m\}$$

$$+ e^{-\delta\Delta} f(R_1 + \Delta(a_1 R_1 - \alpha_1 - \sum_{j=1}^{n} \beta_{1j}R_j - \gamma_1 E), \ldots,$$

$$R_n + \Delta(a_n R_n - \alpha_n - \sum_{j=1}^{n} \beta_{nj}R_j - \gamma_n E))] \qquad (139)$$

Using Taylor's theorem expanding in power of Δ, simplifying and proceeding to the limit as $\Delta \to 0$, we get

$$\delta f(R_1, R_2, \ldots, R_n) = \max_E [\sum_{i=1}^{n} p_i(\alpha_i + \sum_{j=1}^{n} \beta_{ij}R_j + \gamma_i E) - bE^2 - kE - m)$$

$$+ \sum_{i=1}^{n} \{a_i R_i - (\alpha_i + \sum_{j=1}^{n} \beta_{ij}R_j + \gamma_i E)\} \frac{\partial f}{\partial R_i}, \qquad (140)$$

This gives the equations

$$\sum_{i=1}^{n} \left(p_i \gamma_i - \gamma_i \frac{\partial f}{\partial R_i} \right) - 2bE - k = 0 \qquad (141)$$

$$\delta f(R_1, R_2, \ldots, R_n) = \sum_{i=1}^{n} p_i(\alpha_i + \sum_{j=1}^{n} \beta_{ij}R_j - m)$$

$$+ \sum_{i=1}^{n} (a_i R_i - \alpha_i - \sum_{j=1}^{n} \beta_{ij}R_j) \frac{\partial f}{\partial R_i}$$

$$+ \frac{1}{4b}\left(\left(\sum_{i=1}^{n} \gamma_i\left(p_i - \frac{\partial f}{\partial R_i}\right) - k\right)^2 \qquad (142)$$

Equation (142) gives a partial differential equation for determining $\hat{f}$ as a function of $R_1, R_2, \ldots R_n$ and then (141) determines $E(t)$.

EXERCISE 9.3

1. (a) Find the maximum value of $x_1 x_2 \ldots x_n$ subject to $x_1 + x_2 + \ldots + x_n = c$, $x_i \geqslant 0$.

(b) Find the minimum value of $x_1 + x_2 + \ldots + x_n$ subject to $x_1 x_2 \ldots \ldots x_n = d$, $x_i > 0$.

(c) Discuss the relation of duality between these two problems.

2. Find the minimum value of $\sum_{i=1}^{n} p_i$ subject to $- \sum_{i=1}^{n} p_i \ln p_i = \ln \frac{n}{c}$.

3. Write the duals of all problems given in section 9.3.1 and solve them.

4. The energy E_N expanded in compressing a gas in a multistage unit from given initial pressure p to a final pressure P is given by

$$E_N = n\Delta T\left(\frac{\gamma}{\gamma - 1}\right)\left[\left(\frac{p_1}{p}\right)^\alpha + \left(\frac{p_2}{p_1}\right)^\alpha + \ldots + \left(\frac{P}{p_{N-1}}\right)^\alpha - N\right]$$

when $n, R, T, \gamma, \alpha, N$ are constants. Find $p_1, p_2, \ldots, p_{N-1}$ to minimize E_N.

5. Show that the functional equation for maximizing $\sum\limits_{i=1}^{N} g_i(x_i)$ subject to

$\sum\limits_{i=1}^{n} x_i \leqslant c, x_i = 0, 1$ is give by

$$f_N(c) = \max_{x_N=0,1} [g_N(x_N) + f_{N-1}(c - x_N)]$$

$$= \max [g_N(1) + f_{N-1}(c - 1), g_N(0) + f_{N-1}(c)]$$

If $c = 2$, $g_1(x_1) = e^{x_1} - 1$, $g_2(x_2) = e^{-x_2}$, $g_3(x_3) = x_3$, $g_4(x_4) = \ln 2 - xy$, find $f_n(0), f_n(1)$ for $n = 1, 2, 3, 4$.

6. Solve the functional equation (41) viz

$$f_N(x) = \max_{0 \leqslant y \leqslant x} [g(y) + f_{N-1}(a(x - y))], a > 1$$

when $\qquad\qquad g(y) = \sqrt{y}.$

7. Solve the Problem 9.3.2(b) when there is an additional weight constraint

$$\sum\limits_{j=1}^{n} w_j m_j \leqslant w.$$

8. For each of the problem discussed in section 9.2, find $f(x, y)$ by three methods viz. (i) evaluating $I = \int F(x, y, y') \, dx$ (ii) solving equation (123) (iii) solving equation (124), and show that the three approaches lead to the same results.

10

Mathematical Modelling Through Mathematical Programming Maximum Principle and Maximum-Entropy Principle

10.1 MATHEMATICAL MODELLING THROUGH LINEAR PROGRAMMING

Linear programming models are those in which we are required to optimize (maximize or minimize) a linear function of several variables subject to linear inequality and non-negativity constraints on the variables. Thus the general model is

Maximize
$$M = c_1x_1 + c_2x_2 + \cdots + c_nx_n \tag{1}$$

subject to
$$a_{11}x_1 + a_{12}x_2 + \cdots + a_{1n}x_n \leqslant b_1$$
$$a_{21}x_1 + a_{22}x_2 + \cdots + a_{2n}x_n \leqslant b_2$$
$$\cdots \qquad \cdots \qquad \cdots \qquad \cdots \tag{2}$$
$$a_{m1}x_1 + a_{m2}x_2 + \cdots + a_{mn}x_n \leqslant b_m$$

$$x_1 \geqslant 0, \ x_2 \geqslant 0, \ \cdots, \ x_n \geqslant 0 \tag{3}$$

In matrix notation, it can be formulated as

Maximize $C'X$ subject to $AX \leqslant B$ and $X \geqslant 0$, $\qquad$ (4)

where

$$C = \begin{bmatrix} c_1 \\ c_2 \\ \cdot \\ \cdot \\ \cdot \\ c_n \end{bmatrix} \quad X = \begin{bmatrix} x_1 \\ x_2 \\ \cdot \\ \cdot \\ \cdot \\ x_n \end{bmatrix} \quad A = \begin{bmatrix} a_{11} & a_{12} & \cdots & a_{1n} \\ a_{21} & a_{22} & \cdots & a_{2n} \\ \cdots & \cdots & \cdots & \cdots \\ a_{m1} & a_{m2} & \cdots & a_{mn} \end{bmatrix}$$

$$B = \begin{bmatrix} b_1 \\ b_2 \\ \cdot \\ \cdot \\ \cdot \\ b_m \end{bmatrix} \tag{5}$$

10.1.1 Linear Programming Models in Harvesting of Animal Populations

(a) *Constant Population Size Model*

We consider the Leslie model for age-structured population model of section 5.4.2 viz.

$$X(t + 1) = AX(t), \tag{6}$$

where

$$X(t) = \begin{bmatrix} x_1(t) \\ x_2(t) \\ \cdot \\ \cdot \\ \cdot \\ x_n(t) \end{bmatrix},$$

$$A = \begin{bmatrix} -(d_1+m_1) & 0 & 0 & \ldots & b_{p+1} & b_{p+2} & \ldots & b_{p+q} & 0 & \ldots & 0 & 0 \\ m_1 & -(d_2+m_2) & 0 & \ldots & 0 & 0 & \ldots & 0 & 0 & \ldots & 0 & 0 \\ 0 & m_2 & -(d_3+m_3) & \ldots & 0 & 0 & \ldots & 0 & 0 & \ldots & 0 & 0 \\ \cdot\cdot & \cdot\cdot & \cdot\cdot & \ldots & \ldots & \ldots & \ldots & \ldots & \ldots & \ldots & \cdot\cdot \\ 0 & 0 & 0 & \ldots & 0 & 0 & \ldots & 0 & 0 & \ldots & m_{n-1} & -d_n \end{bmatrix}, \tag{7}$$

so that the population vector $X(t)$ at time t becomes $AX(t)$ at time $t + 1$ by a process of natural biological growth.

Now we consider the case when the dominant eigenvalue of the matrix A is greater than unity, so that the population of each species is growing and $AX(t) \geqslant X(t)$. In this case we can harvest the additional net growth of populations and make a profit

$$\begin{aligned} P = &\ p_1[b_{p+1}x_{p+1} + b_{p+2}x_{p+2} + \ldots + b_{p+q}x_{p+q} - (d_1 + m_1 + 1)x_1] \\ &+ p_2[m_1x_1 - (d_2 + m_2 + 1)x_2] + p_3[m_2x_2 - (d_3 + m_3 + 1)x_3] + \ldots \\ &+ p_n[m_{n-1}x_{n-1} - (d_n + 1)x_n), \end{aligned} \tag{8}$$

where $p_1, p_2, \ldots, p_n$ are the profits on units of the n species and $x_1, x_2, \ldots, x_n$ are populations of the n species both at the beginning and at the end of each time interval. Our problem is to maximize the linear function P subject to linear and non-negativity constraints

$$\begin{aligned} b_{p+1}x_{p+1} + b_{p+2}x_{p+2} + \ldots + b_{p+q}x_{p+q} &\geqslant (d_1 + m_1 + 1)\,x_1 \\ m_1x_1 &\geqslant (d_2 + m_2 + 1)\,x_2 \\ m_2x_2 &\geqslant (d_3 + m_3 + 1)\,x_3 \\ \ldots \quad &\ldots \quad \ldots \\ m_{n-1}x_{n=1} &\geqslant (d_n + 1) \quad x_n \end{aligned} \tag{9}$$

$$x_1 \geqslant 0, \quad x_2 \geqslant 0, \ldots, \quad x_n \geqslant 0 \tag{10}$$

This is a linear programming problem and can be solved easily by use of the simplex method. However if $x_1, x_2, \ldots, x_n$ is a solution of the problem then so is $kx_1, kx_2, \ldots, kx_n$ where $k > 0$. As such to get a unique solution, we keep the total population size $x_1 + x_2 + \ldots + x_n$ fixed so that we now seek to maximize P subject to (9), (10) and

$$x_1 + x_2 + \ldots + x_n = K \tag{11}$$

(b) *Growing Population Size Model*

In the above model, we harvested the entire net growth in each time interval, so that the population at the end of each interval is the same as at the beginning of the interval.

We may however decide to harvest only a fraction of the net growth so that the population at the end of each interval of time is more that at its beginning. If the populations at the beginning are $x_1, x_2, \ldots, x_n$, we do the harvesting in such a way that $\lambda x_1, \lambda x_2, \ldots, \lambda x_n$, remain at the end of the interval where $1 \leqslant \lambda < \lambda_0$, where λ_0 is the dominant eigenvalue of the matrix A. If we start with a population vector X, then at the beginnings of successive intervals of time, the populations would be $X, \lambda X, \lambda^2 X, \lambda^3 X, \ldots$ and we follow this policy for N intervals, the profit would be

$$
\begin{aligned}
P' = \frac{\lambda^N - 1}{\lambda - 1} \{ & p_1(b_{p+1}x_{p+1} + b_{p+2}x_{p+2} + \ldots + b_{p+q}x_{p+q} \\
& -(d_1 + m_1 + \lambda)x_1) + p_2(m_1x_1 - (d_2 + m_2 + \lambda)x_2) \\
& +p_3(m_2x_2 - (d_3 + m_3 + \lambda)x_3) + \ldots + p_n(m_{n-1}x_{n-1} - (d_n + \lambda)x_n)\}
\end{aligned}
\tag{12}
$$

and we would seek to maximize P' subject to

$$
\begin{aligned}
b_{p+1}\, x_{p+1} + \ldots + b_{p+q}\, x_{p+q} &\geqslant (d_1 + m_1 + \lambda)\, x_1 \\
m_1\, x_1 \quad\quad &\geqslant (d_2 + m_2 + \lambda)\, x_2 \\
\cdots \quad\quad\quad &\cdots \\
m_{n-1}\, x_{n-1} &\geqslant (d_n + \lambda)x_n
\end{aligned}
\tag{13}
$$

$$x_1 \geqslant 0,\, x_2 \geqslant 0, \ldots, x_n \geqslant 0 \tag{14}$$

$$x_1 + x_2 + \ldots + x_n = K \tag{15}$$

This is again a linear programming model,

(c) *Non-linear density dependent models*

In this case we assume that the birth-rates $b_{p+1}, \ldots, b_{p+q}$, the death rates $d_1, d_2, \ldots, d_n$ and the migration rates $m_1, m_2, \ldots, m_{n-1}$ depend on the total population size.

In case (a), the population size does not change from interval to interval and as such all these rates remain constant so that the linear programming model of subsection (c) continues to hold.

Even in case (b) when the total population sizes are $K, K\lambda, K\lambda^2, \ldots, K\lambda^m$, the birth-death and migration rates change from interval to interval, but are constant in each interval. As such the linear programming model still continues to hold.

10.1.2 Linear Programming Models in Forest Management

Let $x_i(t)$ be the number of trees in the ith height group $(i = 1, 2, \ldots, n)$

at time t. Let g_i be the population of trees in the ith age-group which grow to become trees of the $(i + 1)$th age-group in one period so that a proportion $(1 - g_i)$ of the trees continue to remain in the ith height-group. Let $y_i(t)$ be the number of trees removed from the ith group at the end of this period. Let p_i be the profit on a tree of the ith group. Also let

$$X(t) = \begin{bmatrix} x_1(t) \\ x_2(t) \\ \cdot \\ \cdot \\ \cdot \\ x_n(t) \end{bmatrix}, \quad Y(t) = \begin{bmatrix} y_1(t) \\ y_2(t) \\ \cdot \\ \cdot \\ \cdot \\ y_n(t) \end{bmatrix}, \quad P = \begin{bmatrix} p_1 \\ p_2 \\ \cdot \\ \cdot \\ \cdot \\ p_n \end{bmatrix},$$

$$I_* = \begin{bmatrix} 1 \\ & 1 \\ & & \cdot \\ & & & \cdot \\ & & & & \cdot \\ & & & & & 1 \end{bmatrix} \tag{16}$$

$$G = \begin{bmatrix} 1 - g_1 & 0 & 0 & \ldots & 0 & 0 \\ g_1 & 1 - g_2 & 0 & \ldots & 0 & 0 \\ 0 & g_2 & 1 - g_3 & \ldots & 0 & 0 \\ 0 & 0 & 0 & \ldots & 1 - g_{n-1} & 0 \\ 0 & 0 & 0 & \ldots & g_{n-1} & 1 \end{bmatrix} \tag{17}$$

$$R = \begin{bmatrix} 1 & 1 & 1 & \ldots & 1 & 1 \\ 0 & 0 & 0 & \ldots & 0 & 0 \\ 0 & 0 & 0 & \ldots & 0 & 0 \\ 0 & 0 & 0 & \ldots & 0 & 0 \end{bmatrix} \tag{18}$$

so that

$$I_*' X = x_1 + x_2 + \ldots + x_n, \quad I_* Y = y_1 + y_2 + \ldots + y_n \tag{19}$$

Here $X(t)$ is the population vector, $Y(t)$ is the harvesting vector, P is the profit vector, G is the growth matrix and R is the replacement matrix. For increasing forest wealth, we plant μ times the trees we harvest. The population vector X becomes GX due to growth, is reduced by Y by harvesting, is increased by μRY by planting and should finally be λX so that we get

$$GX - Y + \mu RY = \lambda X \text{ or } (G - \lambda I) X = (I - \mu R)Y \tag{20}$$

Multiplying (20) by I_*', we get

$$I_*' GX - I_*' Y + I_*' \mu RY = \lambda I_*' X \text{ or } X(\mu - 1) I_*' Y = (\lambda - 1) I_*' X \tag{21}$$

which means that the number of additional seedlings planted gives the number of additional trees in the next period.

From (20) and (21)

$$GX - Y + \left[1 + (\lambda - 1) \frac{H \cdot X}{I \cdot Y}\right] RY = \lambda X \tag{22}$$

We take $y_1 = 0$ since it is no use planting new seedlings and then removing them, then (20) gives

$$(1 - g_1)x_1 - y_1$$

$$+ \left[1 + (\lambda + 1) \frac{x_1 + x_2 + \ldots + x_n}{y_1 + y_2 + \ldots + y_n}\right] [y_1 + y_2 + \ldots + y_n] = \lambda x_1$$

$$g_1 x_1 + (1 - g_2) x_2 - y_2 = \lambda x_2$$

$$g_{n-1} x_{n-1} + (1 - g_n) x_n - y_n = \lambda x_n, \tag{23}$$

where $y_1 = 0$, $g_n = 0$. Adding these, we get an identity, so that the last $(n - 1)$ equations are independent. These give the profit function as

$$p_1[g_1 x_1(-\lambda + 1 - g_2)x_2] + p_3[g_2 x_2 + (-\lambda + 1 - g_3)x_3]$$

$$+ \ldots + p_n[g_{n-1} x_{n-1} + (-\lambda + 1)x_n] \tag{24}$$

This has to be maximized subject to

$$g_i x_i + (-\lambda + 1 - g_{i+1})x_{i+1} \geqslant 0, \quad i = 1, 2, \ldots, n - 1 \tag{25}$$

$$x_j \geqslant 0; \quad j = 1, 2, \ldots, n \tag{26}$$

and

$$x_1 + x_2 + \ldots + x_n = s \tag{27}$$

where s is the size of the forest at the beginning of the first period. This size becomes λs, $\lambda^2 s$, $\lambda^3 s$, $\ldots$ at the beginning of second, third, fourth, $\ldots$ periods respectively.

This is the standard linear programming problem.

10.1.3 Transportation and Assignment Models

Let a_1, a_2, $\ldots$, a_m be the number of wagons available at m origins O_1, O_2, $\ldots$, O_m and let b_1, b_2, $\ldots$, b_n be the number of wagons required at n destinations D_1, D_2, $\ldots$, D_n and let

$$\sum_{i=1}^{m} a_i = \sum_{i=1}^{n} b_j \tag{28}$$

Let c_{ij} be the cost of transporting a wagon and let x_{ij} be the number of wagons transported from the ith origin to the jth destination so that the total cost of transportation is

$$C = \sum_{i=1}^{m} \sum_{j=1}^{n} c_{ij} x_{ij} \tag{29}$$

We have to minimize C subject to

$$\sum_{j=1}^{n} x_{ij} = a_i, \quad i = 1, 2, \ldots, m \tag{30}$$

$$\sum_{i=1}^{m} x_{ij} = b_j, \quad j = 1, 2, \ldots, n \tag{31}$$

$$x_{ij} \geqslant 0 \tag{32}$$

This transportation problem is a special case of linear programming problem in which the constraints are equality constraints and x_{ij}'s are non-negative integers. A special transportation algorithm is available for solving this problem.

In the Assignment model, there are n men and n jobs and each man has to be assgined one job. If the ith man is assigned the jth job, the output is c_{ij} so that the total output is

$$\sum_{j=1}^{n} \sum_{i=1}^{m} c_{ij} x_{ij}, \tag{32}$$

where $x_{ij} = 0$ or 1 and the matrix $[x_{ij}]$ has only n non-zero elements, one in every row and one in every column. This is also a special case of linear programming problem for which a special assignment algorithm is available.

10.1.4 Linear Programming Formulation of the Theory of the Firm

Let a_{ij} denote the amount of the ith resource required to produce one unit of the jth commodity $(i = 1, 2, \ldots, m; j = 1, 2, \ldots, n)$. Let x_j be the number of goods of type j produced and c_j be the profit on a unit good of jth type, then the profit made by the firm is

$$Z = c_1 x_1 + c_2 x_2 + \ldots + c_n x_n \tag{33}$$

The amount of the ith resource required for making these goods must be less than or equal to the total quantity b_i of this resource available to the firm so that we have

$$a_{i1} x_1 + a_{i2} x_2 + \ldots + a_{in} x_n \leqslant b_i, \quad i = 1, 2, \ldots, m \tag{34}$$

We have also the non-negativity constraints

$$x_1 \geqslant 0, x_2 \geqslant 0, \ldots, x_n \geqslant 0 \tag{35}$$

so that the linear programming formulation of the theory of the firm is

$$\max Z \text{ subject to (34) and (35)}$$

Given numerical values of a_{ij}'s b_i's and c_j's, we can use the simplex algorithm to find maximum value of Z and the values of $x_1, x_2, \ldots, x_n$ which maximize Z.

According to linear programming theory, if the m constraints (34) are independent, then for the optimal solution only m of the x_j's will be non-zero. It can also be shown that a feasible programme is an optimal programme if and only if it contains a list of included goods such that no excluded goods is more profitable than an equivalent combination in terms of included goods.

The dual of the above linear programming problem is

$$\text{Maximize } z = b_i w_1 + b_2 w_2 + \ldots + b_m w_m \tag{36}$$

subject to

$$a_{1j}w_1 + a_{2j}w_2 + \ldots + a_{mj}w_m \geqslant c_j, \qquad j = 1, 2, \ldots, n \qquad (37)$$

$$w_i \geqslant 0, \qquad i = 1, 2, \ldots, m \qquad (38)$$

Each w_i has the dimension of price and is called the imported or shadow price of the ith resource. It is not the actual price of the ith resource and this price is not given to us and no amount of mathematical manipulation can enable us to get the value of a quantity not given in the problem.

(36) gives the cost of the available resources and we want to minimize this cost. The costs have to be non-negative and the cost of resources required to produce a unit good of jth type is not less than the profit made on it.

It can be shown that the dual of the dual problem is the original or the primal problem.

It the dual problem is easier to solve than the primal problm, we can solve it and then deduce the solution of the primal problem from it.

EXERCISES 10.1

1. For the two matrices

$$A_1 = \begin{bmatrix} 0 & 9 & 12 \\ 1/3 & 0 & 0 \\ 0 & 0 & 1/2 \end{bmatrix}, \; A_2 = \begin{bmatrix} 0 & 3 & 36 \\ 1/3 & 0 & 0 \\ 0 & 1/2 & 0 \end{bmatrix}$$

show that the dominant eigenvalue is $\lambda_0 = 2$ and the corresponding eigenvector is $1/29$ [24, 4, 1]

2. For the matrix A_1 in Ex. 1 and for $p_1 = p_2 = p_3 = 1$, show that the optimal solution is given by $x_1 = 2/3$, $x_2 = 2/9$, $x_3 = 1/9$, $k_1 = 1/5$, $k_2 = 1$ and $k_3 = 0$ and the optimal yield is 8/3.

3. For the matrix A_2 and for $p_1 = 1$, $p_2 = 1$, $p_3 = 100$, show that the optimal solution is given by $x_1 = 3/4$, $x_2 = 1/4$, $x_3 = 0$, $h_1 = 1/3$, $h_2 = 1$ and $h_3 = 0$ and the optimal yield is 14.

4. Develop the optimal harvesting model for continuous-time-discrete-age-scale case of section 3.1.4 show that this also leads to linear programming.

3. Show that the optimal management of forests model for the continuous-time, discrete-age-scale case also leads to a linear programming problem.

6. Show that the dual of the dual problem is the primal problem.

7. Write the duals of the following linear programming problems

(a) Max $z = 2x_1 + 3x_2 + x_3$

subject to the constraints

$$4x_1 + 3x_2 + x_3 = 6$$

$$x_1 + 2x_2 + 5x_3 = 4$$

$$x_1, x_2, x_3 \geqslant 0$$

(b) Min $z = 2x_1 + 3x_2 + 4x_3$

subject to the constraints

$$2x_1 + 3x_2 + 5x_3 \geqslant 2$$
$$2x_1 + x_2 + 7x_3 = 3$$
$$x_1 + 4x_2 + 6x_3 \leqslant 5$$
$$x_1, x_2 \geqslant 0, x_3 \text{ is unrestricted}$$

8. A department head has four subordinates and four tasks have to be performed. Time each man would take to perform each task is given in the following effectiveness matrix. How should the tasks be allocated so as to minimize the total man-hours?

		Subordinates		
	I	II	III	IV
A	8	26	17	11
Tasks B	13	28	4	26
C	38	19	18	15
D	19	26	24	10

Find the time taken for each of the 24 possible assignments.

9. Find some feasible solutions for the transportation problem

	D_1	D_2	D_3	Supply
O_1	6	4	1	10
O_2	8	9	2	8
Demand	6	6	6	18

and find the cost of each.

10.2 MATHEMATICAL MODELLING THROUGH NON-LINEAR PROGRAMMING

10.2.1 Optimal Portfolio Selection : A Quadratic Programming Model

An investor has a unit amount to invest and he can invest it in n securities. The expected return from the ith security is r_i and the variance of this return is σ_i^2. Also the returns from the ith and jth securities are related with a correlation coefficient ρ_{ij} $(i, j = 1, 2, \ldots, n)$. The investor has to find the amounts $x_1, x_2, \ldots, x_n$ which he should invest in the n securities so that his total expected return is maximum and the variance of his return is minimum. If E denotes the expected return and V is the variance of this return, then

$$E = x_1 r_1 + x_2 r_2 + \ldots + x_n r_n = \sum_{i=1}^{n} x_i r_i \tag{39}$$

and
$$V = x_1^2 \sigma_2^1 + x_2^2 \sigma_2^2 + \ldots + x_n^2 \sigma_n^2$$
$$+ 2x_1 x_2 \rho_{12} \sigma_1 \sigma_2 + \ldots + 2x_{n-1} x_n \rho_{n, n-1} \sigma_{n-1} \sigma_n$$

$$= \sum_{i-1}^{n} x_i^2 \sigma_i^2 + 2 \sum_{\substack{j=1 \\ j>1}}^{n} \sum_{i=1}^{n} \rho_{ij} x_i x_j \sigma_i \sigma_j \qquad (40)$$

For every n-tuple $x_1, x_2, \ldots, x_n$ satisfying

$$x_1 \geqslant 0, \; x_2 \geqslant 0, \ldots, x_n \geqslant 0; \; \sum_{i=1}^{n} x_i = 1, \qquad (41)$$

we can find the corresponding E and V and plot the point E, V in the E-V plane. (Figure 10.1)

The set of all these points give a certain region R in the E-V plane.

Every point in this region corresponds to a feasible portfolio

$$(x_1, x_2, \ldots, x_n).$$

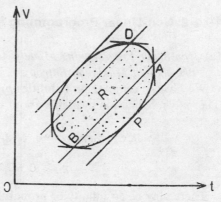

Out of two portfolios giving (E_1, V_1) and (E_2, V_2), we shall prefer the first to the second if

$$E_1 \geqslant E_2 \text{ and } V_1 \leqslant V_2$$

Out of all portfolios corresponding to points on a vertical line parallel to V-axis, we choose the one corresponding to the lowest point since for given

Figure 10.1

E, it gives minimum V. Similarly out of all points on a horizontal line parallel to E-axis, we choose the rightmost point since for a given V, it gives the maximum return. These considerations give points on the curves ABC and BAD respectively. The arc AB is common to both. Thus points on the arc AB give better feasible portfolios than others. However portfolios corresponding to points on the arc AB are not comparable, since out of two portfolios on it if one has greater expected return than the other, it will have variance which will also be greater than the variance of the other portfolio.

The points on the arc AB are said to correspond to points on the mean-variance efficient frontier.

To find points on this frontier, we solve the quadratic programming problem

$$\text{Min } V - \lambda E = \sum_{i=1}^{n} x_i^2 \sigma_i^2 + 2 \sum_{\substack{j=1 \\ j>i}}^{n} \sum_{i=1}^{n} x_i x_j \sigma_i \sigma_j \rho_{ij} - \lambda (x_1 r_1 + \ldots + x_n r_n) \quad (42)$$

subject to

$$x_1 \geqslant 0, \; x_2 \geqslant 0, \ldots, x_n \geqslant 0, \; \sum_{i=1}^{n} x_i = 1 \qquad (43)$$

for each value of $\lambda \geqslant 0$. Since V-λE is a convex function of $x_1, x_2, \ldots, x_n$ the local minimum will also give the global minimum.

When $\lambda = 0$, we have to minimize V and this gives us the portfolio corresponding to point B in Figure 10.1. When $\lambda = \infty$, we have to maximize E and this gives the portfolio corresponding to point A in Figure 10.1.

Other values of λ between 0 and ∞ will corresponding to points on the arc AB between A and B.

If we draw straight lines

$$V - \lambda E = K, \tag{44}$$

for a fixed value of λ and for different values of K, the line corresponding to minimum value of K corresponds to say the point P on the arc AB. For each value of λ, there will be a corresponding point P on the arc AB.

10.2.2 Non-Linear Programming Models in Information Theory

(a) *Non-linear Programming Models Arising from Application of Principle of Maximum Entropy*

We have to estimate probabilities $p_1, p_2, \ldots, p_n$ of n possible outcomes. The only information available about these is that

$$\sum_{i=1}^{n} p_i g_r(x_i) = a_r, \ (r = 0, 1, 2, \ldots, m); \ g_0(x_i) = 1, \ a_0 = 1,$$

$$p_i \geqslant 0 \ \forall i, \ m + 1 < n \tag{45}$$

There may be an infinity of probability distributions satisfying (45) and we have to choose one out of these. According to the principle of maximum entropy, we should choose that one for which the measure of entropy $_l$

$$H_1(P) = - \sum_{i=1}^{n} p_i \ln p_i \quad \text{or} \quad H_2(P) = \frac{1}{1 - \alpha}(\sum_{i=1}^{n} p_i^{\alpha} - 1), \ \alpha \neq 1 \tag{46}$$

is maximum subject to (45). This is obviously a non-linear mathematical programming problem. Thus if we use $H_1(P)$, we can solve it easily by using Lagrangian's method. However if we use $H_2(P)$ as a measure of entropy, we may have to use a standard mathematical programming technique.

(b) *Non-linear Programming Problem Arising from the Application of Principle of Minimum Discrimination Information*

Here in addition to (45), we are also given a priori estimates $q_1, q_2, \ldots, q_n$ for the probabilities, then according to the principle of minimum discrimination information, we choose $p_1, p_2, \ldots, p_n$ by minimising a measure of directed divergence

$$D_1(P:Q) = \sum_{i=1}^{n} p_i \ln \frac{p_i}{q_i} \quad \text{or} \quad D_2(P:Q) = \frac{1}{\alpha - 1}(\sum_{i=1}^{n} p_i^{\alpha} q_i^{1-\alpha} - 1), \ \alpha \neq 1 \tag{47}$$

Both these measures give rise to non-linear mathematical programming problems though if we use the first measure, Lagrange's method is enough to get the solution.

(c) Gain in Information due to Subdivision of Outcomes

If the ith outcome with probability p_i is divided into m_i suboutcomes each with probability p_i/m_i, the gain in information is

$$- \sum_{i=1}^{n} \frac{p_i}{m_i} \ln \frac{p_i}{m_i} - \left(- \sum_{i=1}^{n} p_i \ln p_i \right) = \sum_{i=1}^{n} p_i \ln m_i \tag{48}$$

and we may like to maximise it subject to

$$\sum_{i=1}^{n} m_i c_i = K, \ m_i \text{ a non-negative integer} \tag{49}$$

where c_i is the 'cost' associated with each of the subdivisions of the ith outcome.

This is a non-linear integer programming problem.

If we use $H_2(P)$ as a measure of entropy, the gain in information is

$$\frac{1}{1-\alpha} \left(\sum_{i=1}^{n} m_i \left(\frac{p_i}{m_i}\right)^{\alpha} - 1 \right) - \frac{1}{1-\alpha} \left(\sum_{i=1}^{n} p_i^{\alpha} - 1 \right)$$

$$= \frac{1}{1-\alpha} \left(\sum_{i=1}^{n} m_i^{1-\alpha} p_i^{\alpha} - \sum_{i=1}^{n} p_i^{\alpha} \right) \tag{50}$$

Maximization of (50) subject to (49) again gives a non-linear integer programming problem.

If we use measure of entropy

$$H_3(P) = \frac{1}{1-\alpha} \ln \frac{\sum\limits_{i=1}^{n} p_i^{\alpha+\beta-1}}{\sum\limits_{i=1}^{n} p_i^{\beta}}, \tag{51}$$

the gain in information is

$$\frac{1}{1-\alpha} \ln \frac{\sum\limits_{i=1}^{n} p_i^{\alpha+\beta-1} m_i^{2-\alpha-\beta}}{\sum\limits_{i=1}^{n} p_i^{\beta} m_i^{1-\beta}} - \frac{1}{1-\alpha} \ln \frac{\sum\limits_{i=1}^{n} p_i^{\alpha+\beta-1}}{\sum\limits_{i=1}^{n} p_i^{\beta}} \tag{52}$$

so that we have to maximize

$$\sum_{i=1}^{n} p_i^{\alpha+\beta-1} m_i^{2-\beta-\alpha} / \sum_{i=1}^{n} p_i^{\beta} m_i^{1-\beta} \tag{53}$$

according as $\alpha \gtrless 1$, for variations in m_i's subject to (49). This is a non-linear fractional integer programming problem.

10.2.3 Non-Linear Programming Models Arising from Pollution Control

Polluted water is being discharged into a flowing river at points A_j $(j = 1, 2, \ldots, n)$ and drinking water is being drawn from the river at points $B_i (i = 1, 2, \ldots, m)$. Let

x_j be the quantity of waste water removed (or cleaned) at source j

$f_j(x_j)$ be the cost of removing or cleaning this water

u_j be the upper bound of waste water that can be removed at source j

$A_{ij}(x_j)$ be the improvement in the quality of water at the point i due to removal of waste water x_i at the point j

b_i be the minimum improvement desired at the point i,

then our optimization model is

$$\text{Minimize } z = \sum_{j=1}^{n} f_j(x_j) \tag{54}$$

$$\text{subject to } \sum_{j=1}^{n} A_{ij}(x_j) \geqslant b_i \;\; (i = 1, 2, \ldots, m) \tag{55}$$

$$0 \leqslant x_j \leqslant u_j \qquad (j = 1, 2, \ldots, n) \tag{56}$$

Here $x_1, x_2, \ldots, x_n$ are the decision variables and the function f_j and A_{ij} are supposed to be known. The model as formulated here is a non-linear programming model. If f_i and A_{ij} are linear functions, it becomes a linear programming model.

In the 'equity' model of water management, each source removes the same proportion S of its wastes and we want to minimize S subject to achieving the desired improvements at the m intake points. If a source is already removing more than a fraction S of its waste water, it does not have, to remove more, but if it is removing less than S, it has to make up for the deficiency. Thus if P_j is the fraction being removed at present, it need not remove any additional fraction if $S < P_j$ and it has to remove the additional fraction $S - P_j$ if $S > P_j,$. Thus our model is

Minimize S

subject to

$$\sum_j A_{ij}(x_j) \geqslant b_i \qquad\qquad (i = 1, 2, \ldots, m) \tag{57}$$

$$P_j + \frac{x_j}{u_j} = S \;\; \text{if} \;\; S > P_j \;\; (j = 1, 2, \ldots, n) \tag{58}$$

$$x_j = 0 \qquad \text{if} \;\; S \leqslant P_j \;\; (j = 1, 2, \ldots, n) \tag{59}$$

$$0 \leqslant \frac{x_j}{u_j} \leqslant 1 - P_j \qquad (j = 1, 2, \ldots, n) \tag{60}$$

The cost of treating waste water is

$$\varphi(S) = \sum_{j=1}^{n} f_j(x_j) = \sum_{j=1}^{n}{}' f_j[(S - P_j)u_j] \tag{61}$$

where Σ' denotes summation over only those sources for which $S > P_j$. Since $f_j(x_j)$ is an increasing function of x_j, $\varphi(S)$ decreases and so minimum S implies minimum cost. In fact, one direct method of solving the problem is to continue giving gradually increasing values to S and, for each value of S, calculate x_j and find whether the given constraints are satisfied. The smallest value of S satisfying the constraints gives the desired solution.

EXERCISE 10.2

1. Obtain the solution of the optimal portfolio selection problem where the non-negativity constraints are not imposed i.e. when short sales are allowed.

2. Maximize $- \sum\limits_{i=1}^{n} p_i \ln p_i$ subject to $\sum\limits_{i=1}^{n} p_i = 1, \sum\limits_{i=1}^{n} p_i g_r(x_i) = a_r, (r = 1, 2, \ldots, m)$ and show that the maximum value is a concave function of

$$a_1, a_2, a_3, \ldots, a_m.$$

3. Minimize $\sum\limits_{i=1}^{n} p_i^2$ subject to $\sum\limits_{i=1}^{n} ip_i = m(1 < m < n)$ and $\sum\limits_{i=1}^{n} p_i = 1$ by using Lagrange's method and show that for some value of m, the minimizing p_i's can be negative.

4. Find the minimum value of $\sum\limits_{i=1}^{n} p_i \ln \dfrac{p_i}{q_i}$ subject to $\sum\limits_{i=1}^{n} p_i = 1,$ $\sum\limits_{i=1}^{n} p_i\, g_i(x_i) = a_r (r = 1, \ldots, m)$ and show that this is a convex function of $a_1, a_2, \ldots, a_m.$

5. Use dynamic programming technique to

(i) maximize (48) subject to (49)

(ii) maximize (50) subject to (49)

6. Find the gain in information given by (52) when $m_i = Kp_i.$

10.3 MATHEMATICAL MODELLING THROUGH MAXIMUM PRINCIPLE

10.3.1 Pontryagin's Maximum Principle

This principle enables us to maximize or minimize

$$P = \int_a^b \varphi_0(t, x_1(t), x_2(t), \ldots, x_n(t), h_1(t), \ldots, h_n(t))\, dt \qquad (62)$$

subject to

$$\frac{dx_i}{dt} = \varphi_i(t, x_1, x_2, \ldots, x_n, h_1(t), \ldots, h_n(t)), i = 1, 2, \ldots, n \qquad (63)$$

If we know $h_1(t), h_2(t), \ldots, h_n(t)$ we can solve for $x_1(t), \ldots, x_n(t)$ from equations (63) and then integrate (62) to find P. Thus P is a function of $h_1(t), h_2(t), \ldots, h_n(t)$ and we can choose these control functions in such a manner as to maximize or minimize P.

According to Pontryagin's maximum principle, we form the Hamiltonian function

$$H = \varphi_0 + \sum_{i=1}^{n} \psi_i \varphi_i \qquad (64)$$

where for determining the functions $\psi_1, \psi_2, \ldots, \psi_n$, we have the auxiliary equations

$$\frac{\partial H}{\partial x_i} = -\frac{d\psi_i}{dt}, \quad (i = 1, 2, \ldots, n) \qquad (65)$$

H is a function of $h_1, h_2, \ldots, h_n$ and we choose $h_1, h_2, \ldots, h_n$ to maximize H. This gives us n equations. These equations together with the n equations (63) and the n equation (65) give us $3n$ equations to determine $x_i(t)$, $\psi_i(t)$, $h_i(t)$ $(i = 1, 2, \ldots, n)$.

If $\varphi_0 = 1$, we get $P = t$ and this gives the solution of the time-optimal problem.

10.3.2 Solution of a Simple Time-Optimal Problem

A particle starts from the point at a distance x_0 from the origin on the x-axis with a velocity v_0. It is acted on by a force $u(t)$ along the positive direction of x-axis which is at our disposal, subject to the condition that $|u(t)| \leqslant 1$. The particle is required to reach the origin with zero velocity. We have to determine $u(t)$ so that the time taken in reaching the origin is minimum.

The equation of motion is

$$\frac{d^2x}{dt_2} = u(t) \tag{66}$$

or
$$\frac{dx}{dt} = v(t), \qquad \frac{dv}{dt} = u(t) \tag{67}$$

Equations (64) and (65) then give

$$H = 1 + \psi_1 v(t) + \psi_2 u(t) \tag{68}$$

$$\frac{\partial H}{\partial x} = -\frac{d\psi_1}{dt}, \qquad \frac{\partial H}{\partial v} = -\frac{d\psi_2}{dt} \tag{69}$$

From (68) and (69)

$$0 = -\frac{d\psi_1}{dt}, \quad \psi_1 = -\frac{d\psi_2}{dt} \tag{70}$$

Integrating

$$\psi_1 = c_1, \qquad \psi_2 = c_2 - c_1 t \tag{71}$$

Now we have to maximize H as a function of u when $-1 \leqslant u \leqslant 1$. This gives $u(t) = 1$ whenever ψ_2 is positive and $u(t) = -1$ whenever ψ_2 is negative. As such

$$u(t) = \text{sgn}\ (c_2 - c_1 t) \tag{72}$$

Integrating

$$x = \frac{1}{2}v^2 + A \text{ when } u = 1, \quad x = -\frac{1}{2}v^2 + B \text{ when } u = -1 \tag{73}$$

These represent two sets of parabolas in the $x - v$ plane. When v is positive, x increases and when v is negative x decreases. Thus we get the two sets of directed parabolas I and II as shown in Figure 10.2. Through each point of the $x - v$ plane, there passes one parabola of each family. Thus through x_0, v_0 there will be two parabolas, one of which will take the phase point completely away from the origin. As such we let the phase point move along the other parabola till it meets the shaded arc of the parabola of the other family and then the phase point moves along it to the origin.

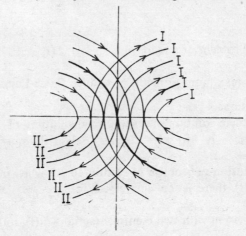

Figure 10.2

Thus we apply the force $u(t) = 1$ in the beginning and $u(t) = -1$ after some time or we apply $u(t) = -1$ first and then $u(t) = 1$ afterwards and which option we use depends on the point we start with in the phase plane.

10.3.3 Optimal Harvesting of Animal Populations

We consider again the problem discussed in section 9.2.5 of maximizing

$$P = \int_0^\infty e^{-\delta t}(p - c(x))h(t)\, dt \tag{74}$$

subject to
$$\frac{dx}{dt} = F(x) - h(t) \tag{75}$$

Here
$$H = e^{-\delta t}(p - c(x))h(t) + \psi(t)[F(x) - h(t)] \tag{76}$$

Equation (65) gives

$$e^{-\delta t}(-c'(x)\,h(t)) + \psi(t)F'(x) = -\frac{d\psi}{dt} \tag{77}$$

Also for maximizing H as a function of h, we find that

$$h = M \text{ if } e^{-\delta t}(p - c(x)) - \psi(t) > 0 \tag{78}$$
$$h = 0 \text{ if } e^{-\delta t}(p - c(x)) - \psi(t) < 0, \tag{79}$$

where M is the maximum permissible rate of harvesting. This gives us the bang-bang solution i.e. either do no harvesting or do maximum rate harvesting. If

$$e^{-\delta t}(p - c(x)) - \psi(t) = 0, \tag{80}$$

we get

$$\psi(t)F'(x) = -\frac{d\psi}{dt} \tag{81}$$

and this gives the steady-state solution in which $h(t)$ and $F(x)$ are equal and $x(t)$ is a constant. This is the solution found earlier by using calculus of variations.

EXERCISE 10.3

1. Consider the problem of maximizing $I = \int_a^b F(t, x, h(t)) \, dt$ subject to $\frac{dx}{dt} = h(t)$. Use the maximum principle to deduce Euler-Lagrange equation of calculus of variations.

2 Find the complete solution of time-optimal solution of section 10.3.2 when (i) $x_0 = 1, v_0 = 1$, (ii) $x_0 = 1, v_0 = -1$, (iii) $x_0 = -1, v_0 = 1$, (iv) $x_0 = -1, v_0 = -1$.

3. Obtain the solution of problem of section 10.3.2 when in addition to the control force $u(t)$, there is an impressed force $-x(t)$ acting on the particle.

4. Consider the system with two control functions $u_1(t)$, $u_2(t)$

$$\frac{dx}{dt} = v + u_1(t), \frac{dv}{dt} = -x + u_2(t)$$

where $|u_1(t)| \leqslant 1$, $|u_2(t)| \leqslant 1$. Solve the problem of reaching the origin with zero velocity in minimum time.

5. Extend the problem of section 10.3.3 to the harvesting of two species of predator and prey animals.

6. Discuss the relationship between calculus of variations, maximum principle and dynamic programming. Are these equivalent? For which types of models will you use each?

10.4 MATHEMATICAL MODELLING THROUGH THE USE OF PRINCIPLE OF MAXIMUM ENTROPY

We have already discussed the Maximum-Entropy Principle and some of its applications in Section 9.2. We give below some more mathematical models illustrating the power of this important principle which is ideally suited for those situations where the information given is incomplete.

10.4.1 Maxwell-Boltzmann Distribution in Statistical Mechanics

We want to estimate the probabilities of a particle being in the n energy levels $\epsilon_1, \epsilon_2, \ldots, \epsilon_n$ when the only knowledge available about the system is the value of the average energy of the system. According to the principle of maximum-entropy, to get the most unbiased estimates of probabilities, we maximize the entropy

$$S = - \sum_{i=1}^n p_i \ln p_i \tag{82}$$

subject to

$$\sum_{i=1}^n p_i = 1, \sum_{i=1}^n p_i \epsilon_i = \bar{\epsilon}, \quad p_i \geqslant 0 \tag{83}$$

Using Lagrange's method, this gives

$$p_i = e^{-\mu \epsilon_i} / \sum_{i=1}^{n} e^{-\mu \epsilon_i}, \quad i = 1, 2, \ldots, n \tag{84}$$

where the Lagrange's multiplier μ is determined by using (83) so that

$$\sum_{i=1}^{n} \epsilon_i e^{-\mu \epsilon_i} / \sum_{i=1}^{n} e^{-\mu \epsilon_i} = \bar{\epsilon} \tag{85}$$

The probability distributions (84) is known as Maxwell-Boltzmann distribution. From (85)

$$\frac{d\bar{\epsilon}}{d\mu} = \frac{-(\sum_{i=1}^{n} e^{-\mu \epsilon_i})(\sum_{i=1}^{n} \epsilon_i^2 e^{-\mu \epsilon_i}) + (\sum_{i=1}^{n} \epsilon_i e^{-\mu \epsilon_i})^2}{(\sum_{i=1}^{n} e^{-\mu \epsilon_i})^2} \tag{86}$$

By using Cauchy-Schwarz inequality $\sum_{i=1}^{n} a_i^2 \sum_{i=1}^{n} b_i^2 \geqslant (\sum_{i=1}^{n} a_i b_i)^2$, it is easily seen that the numerator of the RHS of (86) $\leqslant 0$ so that

$$d\bar{\epsilon}/d\mu \leqslant 0 \tag{87}$$

Thus μ is a monotonic decreasing function of the average energy $\bar{\epsilon}$ and if we put

$$\mu = \frac{1}{kT}, \tag{88}$$

then T is a monotonic increasing function of $\bar{\epsilon}$. We *define* T as the thermodynamic temperature of the system.

Substituting from (84) in (82), we get the value S_{max} of the maximum entropy as

$$S_{max} = - \sum_{i=1}^{n} p_i(-\mu \epsilon_i - \ln \sum_{i=1}^{n} e^{-\mu \epsilon_i})$$

$$= \mu \bar{\epsilon} + \ln \sum_{i=1}^{n} e^{\mu \epsilon_i}, \tag{89}$$

so that

$$dS_{max} = \mu \, d\bar{\epsilon} + \bar{\epsilon} d\mu + \frac{\sum_{i=1}^{n} e^{-\mu \epsilon_i}(-\mu d\epsilon_i - d\mu \epsilon_i)}{\sum_{i=1}^{n} e^{-\mu \epsilon_i}}$$

$$= \mu(d\bar{\epsilon} - \sum_{i=1}^{n} p_i \, d\epsilon_i) + d\mu \left(\bar{\epsilon} - \frac{\sum_{i=1}^{n} \epsilon_i e^{-\mu \epsilon_i}}{\sum_{i=1}^{n} e^{-\mu \epsilon_i}} \right)$$

$$= \mu \sum_{i=1}^{n} \epsilon_i \, dp_i, \tag{90}$$

on making use of (83) and (85). Again from (83)

$$d\bar{\epsilon} = \sum_{i=1}^{n} p_i \, d\epsilon_i + \sum_{i=1}^{n} \epsilon_i \, dp_i \tag{91}$$

The first term on the right is due to change in energies and is called the work effect and is denoted by $-\Delta W$. The second term is due to changes in probabilities of various states and is called the heat effect and is denoted by ΔH, so that

$$d\bar{\epsilon} = -\Delta W + \Delta H, \tag{92}$$

so that (90) gives

$$dS_{\max} = \mu\Delta H = \frac{\Delta H}{kT} \tag{93}$$

$S_{\max}$ is defined on the thermodynamic entropy. Thus thermodynamic entropy is the maximum possible information-theoretic entropy of a system having a given average energy.

Thus our model defines in a very natural manner temperature, work effect, heat effect and thermodynamic entropy. From (92) we get

$$\oint (\Delta H - \Delta W) = 0 \tag{94}$$

If $\epsilon_1 < \epsilon_2 < \ldots < \epsilon_n$, then when $T \to 0$, $\mu \to \infty$ and from (84) $p_1 = 1$ and all other probabilities tend to zero so that all the particles tend to be in the lowest energy state.

In fact all the four laws of thermodynamic can be obtained by combining the concepts of entropy from information theory and the concept of energy from mechanics.

10.4.2 Bose-Einstein, Fermi-Dirac and Intermediate Statistics Distributions

(a) Bose-Einstein Distribution

In the last subsection, we assumed the knowledge of only the average energy $\bar{\epsilon}$ of the system. Now we assume that we know in addition the expected number of particles in the whole system.

Let p_{ij} be the probability of there being j particles in the ith energy state, then we are given that

(i) $$\sum_{j=0}^{\infty} p_{ij} = 1, \qquad i = 1, 2, \ldots, n, \tag{95}$$

since it is certain that there will be k particles in the ith energy level where k may be 0 or 1 or 2 or 3 . . .

(ii) $$\sum_{i=1}^{\infty} \sum_{j=0}^{n} jp_{ij} = N, \tag{96}$$

since the expected number of particles in the system is supposed to be known

(iii) $$\sum_{i=1}^{\infty} \epsilon_i \sum_{j=0}^{n} jp_{ij} = N\bar{\epsilon} \tag{97}$$

since the average energy of the system is supposed to be known. Now (95), (96), (97) give only $(n + 2)$ pieces of information whereas we have to determine an infinity of p_{ij}'s $(i = 1, 2, \ldots, n; j = 0, 1, 2, \ldots, \infty)$. The equations are obviously not sufficient to determine all p_{ij}'s uniquely. Thus here we have a case of mathematical modelling with partial information only and we appeal to the principle of maximum entropy. We maximize

$$S = - \sum_{i=1}^{n} \sum_{j=0}^{\infty} p_{ij} \ln p_{ij} \tag{98}$$

subject to (95), (96) and (97) to get

$$p_{ij} = a_i e^{-(\lambda + \mu \epsilon_i)j}, \tag{99}$$

Using (95) we get

$$p_{ij} = (1 - e^{-(\lambda + \mu \epsilon_i)}) e^{-(\lambda + \mu \epsilon_i)j} \tag{100}$$

Let $\bar{n}_i$ denote the expected number of particles in the ith energy level, then

$$\bar{n}_i = \sum_{j=0}^{\infty} j p_{ij} = (1 - e^{-(\lambda + \mu \epsilon_i)}) \sum_{j=0}^{\infty} j e^{-(\lambda + \mu \epsilon_i)j} \tag{101}$$

or

$$\bar{n}_i = \frac{e^{-(\lambda + \mu \epsilon_i)}}{1 - e^{-(\lambda + \mu \epsilon_i)}} = \frac{1}{e^{(\lambda + \mu \epsilon_i)} - 1}, \quad i = 1, 2, \ldots, n \tag{102}$$

where λ, μ are determined by using (96), (97) i.e. from

$$\sum_{i=1}^{n} \bar{n}_i = N, \quad \sum_{i=1}^{n} \bar{n}_i \epsilon_i = N\bar{\epsilon} \tag{103}$$

Distribution (102) is known as Bose-Einstein distribution. It gives the expected number of particles in each energy level when the total expected number of particles and the total expected energy of the system are known.

(b) Fermi-Dirac Distribution

In the above discussion, we assumed that the number of particles in any energy state can be any number from 0 to ∞. There are however some particles which have the property that if one particle is already in any energy level, then no other particle can be in that energy level, so that j can take only two values 0 and 1. Here then we maximize

$$- \sum_{i=1}^{n} \sum_{j=0}^{1} p_{ij} \ln p_{ij} \tag{104}$$

subject to

$$\sum_{j=0}^{1} p_{ij} = 1, i = 1, 2, \ldots, n; \quad \sum_{i=1}^{n} \sum_{j=0}^{1} p_{ij} = N;$$

$$\sum_{i=1}^{n} \epsilon_i \sum_{j=0}^{1} j p_{ij} = N\bar{\epsilon} \tag{105}$$

to get $\quad p_{ij} = (1 + e^{-(\lambda + \mu \epsilon_i)})^{-1} e^{-(\lambda + \mu \epsilon_i)j} \tag{106}$

and
$$\bar{n}_i = \sum_{j=0}^{1} jp_{ij} = p_{i1} = \frac{1}{e^{(\lambda+\mu\epsilon_i)} + 1},$$
$$i = 1, 2, \ldots, n \tag{107}$$

where λ, μ are still obtained by using (103). Distribution (106) is known as Fermi-Dirac distribution.

Neither (102) nor (107) gives a probability distribution, though in either case $\bar{n}_i/N$ can be regarded as giving a probability distribution as $\sum_{i=1}^{n} \bar{n}_i = N$.

(c) Intermediate Statistics Distributions

All particles in Nature are Bosons or Fermions i.e. these follow either Bose-Einstein or Fermi-Dirac distribution. However we can theoretically consider the possibility of there being a maximum of m_i particles in the ith energy level. Proceeding as above we get

$$p_{ij} = e^{-(\lambda+\mu\epsilon_i)j} / \sum_{j=1}^{m_i} e^{-(\lambda+\mu\epsilon_i)j} = x_i^j / \sum_{j=1}^{m_i} x_i^j ; \tag{108}$$

where
$$x_i = e^{-(\lambda+\mu\epsilon_i)}, \tag{109}$$

so that

$$n_i = \sum_{j=0}^{m_i} jp_{ij} = \sum_{j=0}^{m_i} jx_i^j / \sum_{j=0}^{m_i} x_i^j$$

$$= \frac{x_i + 2x_i^2 + 3x_i^3 + \ldots + m_i x_i^{m_i}}{1 + x_i + x_i^2 + \ldots + x_i^{m_i}}, i = 1, 2, \ldots, n \tag{110}$$

This is called Intermediate Statistics Distribution or Gentile Statistics Distribution. This was studied in Physics, not because it occurred in nature, but because it could help in understanding the transition from Bose-Einstein to Fermi-Dirac distribution and vice-versa.

However it can arise in social and economic situations. Some typical models giving rise to this distribution would be:

(i) Let p_{ij} be the probability of j beds being occupied in the ith ward of a hospital and let m_i be the number of beds in the ith ward. Let the expected occupancy and the expected income in the hospital be given (where incomes from different wards are different), we can estimate the expected occupancy of each ward.

(ii) Let p_{ij} be the probability of j orders of size i being received by a firm. Let the expected number of orders and expected number of items ordered be given, then we can estimate the expected number of order of each size.

(iii) Let p_{ij} be the probability of j accounts of size i in a bank. Let the total number of accounts and the total deposits in the bank be known, then we can estimate the number of accounts of each size.

(iv) Let p_{ij} be the probability of a firm selling j items of price c_i. Let the total number of items sold and the total sale price be known, then we can estimate the number of items of each type sold.

(v) Let p_{ij} be the probability of a country having j cities with population N_i. Let the total number of cities and the total population be given, then we can estimate the number of cities with a given population.

(iv) In (i) we can consider hotels in place of hospitals.

10.4.3 Econodynamics: An Information-theoretic Model for Economics

Let $c_1, c_2, \ldots, c_n$ be the costs of travel from n colonies to central business district and let the average travel budget $\sum\limits_{i=1}^{n} p_i c_i = \bar{c}$ be known, then to estimate the proportions $p_1, p_2, \ldots, p_n$ of the population living in these colonies, we maximize the entropy $-\sum\limits_{i=1}^{n} p_i \ln p_i$ subject to $\sum\limits_{i=1}^{n} p_i = 1$ and $\sum\limits_{i=1}^{n} p_i c_i = \bar{c}$ to get

$$p_i = \exp(-\mu c_i)\Big/ \sum_{i=1}^{n} \exp(-\mu c_i), \quad i = 1, 2, \ldots, n \tag{111}$$

which is Maxwell-Boltzmann distribution and we can proceed as in Section 10.4.1 to define an economic temperature $T = 1/\mu\bar{c}$, an economic heat $\Delta H = \sum\limits_{i=1}^{n} c_i dp_i$ and an economic entropy by $dS_{max} = \sum\limits_{i=1}^{n} c_i dp_i / T$. From (89)

$$S_{max} = \mu\bar{c} + \ln \sum_{i=1}^{n} \exp(-\mu c_i) \tag{112}$$

Keeping $c_1, c_2, \ldots, c_n$ fixed, S_{max} is a function of $\bar{c}$ and

$$\frac{dS_{max}}{d\bar{c}} = \mu + \bar{c}\frac{d\mu}{dc} + \frac{\sum\limits_{i=1}^{n} \exp(-\mu c_i)\left(-c_i\frac{d\mu}{d\bar{c}}\right)}{\sum\limits_{i=1}^{n} \exp(-\mu c_i)}$$

$$= \mu + \bar{c}\frac{d\mu}{dc} - \bar{c}\frac{d\mu}{d\bar{c}} = \mu \tag{113}$$

$$\frac{d^2 S_{max}}{d\bar{c}^2} = \frac{d\mu}{d\bar{c}}$$

$$= \frac{\left(\sum\limits_{i=1}^{n} \exp(-\mu c_i)\right)^2}{\left(\sum\limits_{i=1}^{n} \mu_i \exp(-\mu c_i)\right)^2 - \left(\sum\limits_{i=1}^{n} \exp(-\mu c_i)\right)\left(\sum\limits_{i=1}^{n} c_i^2 \exp(-\mu c_i)\right)}$$

$$\leqslant 0. \tag{114}$$

so that S_{max} is a concave function of $\bar{c}$. If we arrange $c_1, c_2, \ldots, c_n$ in ascending order then when

$$\bar{c} = c_1, p_1 = 1, p_2 = 0, \ldots, p_n = 0 \text{ and } S = 0 \tag{115}$$

when $\quad \bar{c} = c_n, \ p_1 = 0, p_2 = 0, \ldots, p_n = 1 \text{ and } S = 0 \tag{116}$

and $\quad \bar{c} = c^* = \dfrac{1}{n}(c_1 + c_2 + \ldots + c_n), p_1 = p_2 = \ldots = p_n = \dfrac{1}{n}$

$$\text{and } S = \ln n \tag{117}$$

Figure 10.3 gives the graph of S_{max} against $\bar{c}$. As $\bar{c}$ increases from c_1 to c^*, S_{max} increases from 0 to $\ln n$ and as $\bar{c}$ increases from c^* to c_n, S_{max} decreases from $\ln n$ to 0.

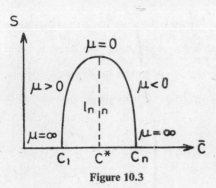

S

Figure 10.3

If the budget allowed is $\bar{c}$ and $\bar{c} > c^*$ and we insist on spending the whole budget, we get $S_{max} < \ln n$, so that by spending a smaller amount c^*, we can get a larger entropy. Thus a more realistic formulation of our model would be

$$\text{Max.} -\sum_{i=1}^{n} p_i \ln p_i$$

subject to $\qquad \sum_{i=1}^{n} p_i = 1, \ \sum_{i=1}^{n} p_i c_i \leqslant \bar{c} \tag{118}$

In this case μ and T would always be positive and only the left-hand part of the $S_{max} - \bar{c}$ curve would be meaningful.

Thus the population tends to distribute itself uniformly over the n colonies subject to the cost constraint. If $\bar{c} > c^*$, the cost constraint becomes ineffective and the population distributes itself completely uniformly over the n colonies.

If we take the energy constraint, we get thermodynamic laws and if we take the cost constraint, we get economodynamic laws. Thus mathematical modelling through maximum-entropy principle shows that we should not be surprised in finding similarities between laws of thermodynamics and economodynamics.

10.4.4 Gravity Model for Transportation Problem in Urban and Regional Planning

There are m residential colonies $A_1, A_2, \ldots, A_m$ in which $a_1, a_2, \ldots, a_m$ office workers live and there are n offices $B_1, B_2, \ldots, B_n$ in which $b_1, b_2, \ldots, b_n$ workers work so that

$$\sum_{i=1}^{m} a_i = \sum_{j=1}^{n} b_j = T, \tag{119}$$

where T is the total number of office workers in all the colonies. Let T_{ij} be the number of workers travelling from the ith residential colony to the jth office so that

$$\sum_{j=1}^{n} T_{ij} = a_i, \ \sum_{i=1}^{m} T_{ij} = b_j \tag{120}$$

(119) and (120) give $m + n - 1$ equations to determine mn unknown quantities T_{ij}'s $(i = 1, 2, \ldots, m; j = 1, 2, \ldots, n)$. Obviously these equations are not sufficient to determine T_{ij}'s uniquely and we appeal to the principle of maximum entropy. We maximize the entropy

$$S = -\sum_{i=1}^{m} \sum_{j=1}^{n} \frac{T_{ij}}{T} \ln \frac{T_{ij}}{T} \tag{121}$$

subject to (119) and (120) and the cost constraint.

$$\sum_{i=1}^{m} \sum_{j=1}^{n} T_{ij} c_{ij} = \bar{c} \tag{122}$$

to get

$$T_{ij} = A_i B_j a_i b_j \exp\left(-\nu c_{ij}\right) \tag{123}$$

The constants $A_i (i = 1, 2, \ldots, m)$, $B_j (j = 1, 2, \ldots, n)$, and ν can be determined by using (119), (120) and (122).

This method is called gravity model of transportation since (123) was deduced by starting from

$$T_{ij} = K \frac{a_i b_j}{c_{ij}^2} \tag{124}$$

on the analogy of Newton's laws of gravitation and then modifying it empirically over a period of thirty years to make it consistent mathematically and with observations. The formula which took thirty years to develop empirically and by trial and error could be deduced in a straight forward manner by using mathematical modelling through the maximum entropy principle.

Another advantage of the mathematical modelling approach over the empirical approach is that having obtained (123) we can generalise it theoretically to the cases when

(i) we want to take into account different modes of transport with different costs

(ii) we want to take into account travel by transit points.

(iii) even $b_1, b_2, \ldots, b_n$ are not given, only n is specified and so on.

10.4.5 Computerised Tomography

To find where a blood clot or a cellular growth has taken place in the brain, we send a large number of photon beams across a section of the brain. Let $f(x, y)$ be the coefficient of absorption at the point (x, y), I_0 be the intensity at the entry point and I_1 be the intensity at the exit point, then

$$\ln \frac{I_1}{I_0} = -\int_L f(x, y) \, dS \tag{125}$$

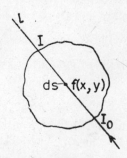

Figure 10.4

We can measure I_0 and I_1 and thus find value of the line integral. If we can find *all* possible line integrals, then we can invert these by using, Radon transform technique to find $f(x, y)$. However in practice we cannot carry out an infinite number of measurements, though we can find a very large number, say 10^5, line integrals. We cannot therefore determine $f(x, y)$ with complete certainty and there is some uncertainty. We accordingly use the principle of maximum entropy and choose $f(x, y)$ to maximize

$$-\iint f(x, y) \ln f(x, y)\, dx dy \quad \text{or} \quad \iint \ln f(x, y)\, dx dy \qquad (126)$$

subject to

$$\int_{L_i} f(x, y)\, ds = a_i, \quad i = 1, 2, \ldots, 10^5 \qquad (127)$$

In practice, we discretise both (126) and (127) by dividing the slice (Tomos) into a very large number of cells.

EXERCISE 10.4

1. When will (87) reduce to an equality?
2. What does (84) approach as $\mu \to 0$, $T \to \infty$?
3. The larger the temperature, the more uniform is the distribution of particles in the energy states. In what sense is this statement true?
4. If $\epsilon_i = i$, $n = 10$, find Maxwell-Boltzmann distributions if $\bar{\epsilon} = 3$ or 5 or 7.
5. Let p_{ijk} be the probability of a firm receiving j orders of size i and cost per item k. Given the expected number of orders, the expected number of items ordered and the expected values of the items ordered, estimate p_{ijk}, $\bar{n}_{ij}$, $\bar{n}_{ik}$, $\bar{n}_{jk}$, $\bar{n}_i$, $\bar{n}_j$, $\bar{n}_k$ and interpret your results.
6. Show that Intermediate Statistics distributions approaches Fermi-Dirac distribution if each $m_i = 1$ and it approaches Bose-Einstein distribution if each $m_i \to \infty$.
7. Discuss in detail the six examples of occurrence of Intermediate Statistics distribution given in section 10.4.2(c) and give four more examples of the same type.
8. Let $f(r)$ be the population density at a distance r from the central business district and let $c(r)$ be the cost of travel. Estimate $f(r)$ by maximizing $-\int_0^\infty f(r) \ln f(r)\, dr$ subject to $\int_0^\infty f(r)\, dr = 1$, $\int_0^\infty c(r) f(r)\, dr = c$. Take some plausible cost functions and deduce the corresponding density functions.

Appendix I

MATHEMATICAL MODELS DISCUSSED IN THE BOOK

(The number within bracket gives the section in which the corresponding model occurs)

A: Mathematical Models in Physical Sciences and Engineering

Astronomy	: Length of day (1.7), Duration of Twilight (1.7)
Celestial Machanics	: Mass of the Earth (1.1), Temperature of the Sun (1.1) Planetary Motion (1.5), Radius of Earth (1.6), Distance of Moon (1.7), Distances of Stars (1.7), Kepler's laws of planetary motion (4.1.5)
Chemistry	: Rate of Dissolution (2.3.3), Laws of Mass Action (2.3.4)
Defence	: Richardson's arms race model (3.5.2), Lanchester's Combat Model (3.5.3), External Ballistics of gun shells (3.6.3), Pursuit of Objects (4.4.2)
Diffusion	: Simple Model (2.2.7), Compartment Model (2.4.1), Succession of compartments (2.4.3)
Dynamics	: Simple Harmonic Motion (2.5.1), Motion in a resisting medium (2.5.2), Motion of a Projectile (3.6.2), Motion Under Central Forces (4.1.1-4.1.4), Circular Motion on wires (4.2.1, 4.2.2), Rectilinear Motion (4.3.1), Modelling through Calculus of Variations (9.2.4), Time Optimal Problem (10.3.2)
Elasticity	: Vibrating Strings (6.3.2, 6.4.3), Vibrating membrane (6.3.2), Integral equation for elasticity (8.2.1)
Electricity	: Electrical Circuits (4.3.2), Electrical Networks and Kirchoff's laws (7.5.1)
Fluid Dynamics	: Equation of Continuity (6.2.1), Euler's equations of motion (6.3.1), Potential gas flows (8.2.1)
Heat	: Decreasing Temperature (2.2.6), Equation of Continuity (6.2.2)
Light	: Laws of reflection (1.5, 1.8), Laws of refraction (1.8), Wave Motion (6.3.4), Optics through Calculus of Variations (9.2.6)
Radioactivity	: Radioactive Decay (2.2.5), Carbon Dating (2.2.5)

Space Flight : Periodic time of satellites (1.1), Motion of a satellite (1.6), Motion of rocket (2.5.3), Circular and elliptic motions of satellites (4.2.3-4.2.4)

Statistical Mechanics : Maxwell-Boltzmann Distribution (10.4.1), Bose-Einstein and Fermi-Dirac Distributions, Intermediate Statistical Distributions (10.4.2)

Surveying : Height of a tower (1.1), Width of a river (1.1)

Technology : Gun with best performance (1.1) life span of a light bulb (1.1), Solar heater (1.5), Parabolic Mirrors (1.5) Elliptic Sound Gallery (1.5), Computerised Tomography (10.4.5)

Thermodynamics : (10.4.1).

B. Mathematical Models in Life Sciences and Medicine

Agriculture : Yield of Wheat (1.1), Food Webs (7.2.5)

Ecology : Pollution of Water (1.2), Pollution Control (6.2.3), Food Webs (7.2.8)

Epidemics : Spread of infectious diseases (2.3.2), Simple epidemic models (3.2.1), Susceptible-Infectives-Susceptible Models (3.2.2, 3.2.3), Epidemic Models with removals (3.2.5, 3.2.6), Stochastic epidemic models (6.5.2-6.5.2)

Forests : Linear Programming in Forest Management (10.1.2)

Fisheries : Population of fish in a pond (1.2), Optimal harvesting (10.3.3), Linear Programming (10.1.2).

Genetics : Modelling Through Difference Equations (5.4.3), Genetic Graphs (7.2.3)

Hospitals : Number of beds (10.4.2)

Medicine : Diffusion of Medicine (2.4.2), Model for Diabetes Mellitus (3.5.1)

Population Dynamics : Estimating the population of a country (1.1.1), Population Growth Models (2.2.1), Effects of immigration and emigration (2.2.3), Logistic Model of population growth (2.3.1), Prey-Predator Models (3.1.1) Competition Models (3.1.2), Multispecies Models (3.1.3), Age-structured population Models (3.1.4, 5.4.2), Non-linear difference equation models (5.4.1), Integral equation approach (5.2.3), Delay-Differential and Differential-Difference equations approaches (8.3, 8.3.1-8.3.8), Birth-Death-Emigration Model (6.5.3)

Physiology : Volume of blood in human body (1.1)

C. Mathematical Models in Economic and Social Sciences

Actuarial Science : Estimating the amount of insurance claims (1.1), Difference Equations Models (6.2.9)

Business Models : (1.8), (10.4.2)

Bioeconomics : Optimal harvesting of animal populations (10.3.3), Mathematical modelling through calculus of variations (9.2.4)

Compound Interest : Interest compounded continuously (2.2.4), Formulae for Compound Interest (8.1.3)

Economic Models : Changes in price of a Commodity (2.2.8), Domar Model (3.4.1), Domar Debt Models (3.4.2, 3.4.3), Allen's speculative Model (3.4.4), Samuelson's Investment Models (3.4.5, 3.4.6), Stability of market equilibrium (3.4.7), Leontief models (3.4.8), Phillips Stabilization Model (4.3.3). Harrod Model (5.2.9), Cobweb Model (5.3.2), Samuelson's interaction models (5.3.3)

Economodynamics : Information-Theoretic approach (10.4.3)

International Trade : (3.5.4)

Portfolio Analysis : Optimal portfolio selection (10.2.1)

Social Sciences : Growth of Science and Scientists (2.2.2), Spread of technological innovations (2.3.2), Senior-subordinate relations (7.2.4), Detection of Cliques (7.2.8), Communication networks (7.2.6, 7.4.1, 7.4.3)

Traffic Problem : Traffic flow (6.2.3), general discussion (6.6.3), Wave propagation (6.6.5), one-way traffic problem (7.2.2)

Urban and Regional Planning : Gravity Models for Transportation (10.4.4); Sizes of cities (10.4.2)

D. Mathematical Models in Management Sciences

Allocation Problem 9.3.2
Assignment Problem 10.1.3
Inventory Theory 1.8
Location of Facilities 1.2
Replacement Theory : Optimum time for replacement
Transportation Problem : 10.1.3

E. Mathematical Models in Information Sciences

Entropy : Entropy of a probability distribution (8.1.4), Maximum Entropy (1.6), Maximum value (1.6), Shannon's inequality (1.6)

Maximum Entropy Models : Maximum entropy Distributions (9.2.2), Statistical Mechanics distributions (10.4.1), (10.4.2), Economodynamics (10.4.3), Gravity Models (10.4.4), Computerised Tomography (10.4.5)

Nonlinear   : In information sciences (10.2.2), Portfolio Analysis
Programming   (6.2.1)

F. Optimization Models

Linear Programm- : Harvesting of animal populations (10.1.1), Forest
ing Models   Management (10.1.2), Transportation and Assignment
                    Models (10.1.3), Theory of the Firm (10.1.4)
Nonlinear   : Optimal Portfolio Selection (10.2.1), Information
Programming   Theory (10.2.2), Pollution Control (10.2.3)
Models
Dynamic   : Allocation problems (9.3.2), Cargoloading problems
Programming   (9.3.2), Reliability of multicomponent systems (9.3.2),
Models   Farmer's problem (9.3.2), Purchase problem (9.3.2)
                    Transportation problem (9.3.5), Calculus of Variations
                    (9.3.2)
Maximum Principle: Time-optimal problem (10.3.2), Optimal harvesting of
                    animal populations.

G. Geometrical Models

Area of a Rectangle (8.1.2)
Geometrical Models Through
Calculus of Variations (9.2.3)
Minimal Surfaces (6.4.2)
Orthogonal Trajectories (2.6.2)
Planar Graphs (7.5.4)
Parallelopiped with given Perimeter and Maximum Volume (1.8)
Regular Solids (7.3.6)
Simple Geometrical Models (2.6.1)
Triangle with a given Perimeter and Maximum Area (1.8)

Appendix II

SUPPLEMENTARY BIBLIOGRAPHY

This bibliography supplements the bibliography given on pages 28-29. It contains books on mathematical modelling as well as on mathematical techniques needed for obtaining the solutions of the mathematical models.

32. J. Aczel "Lectures on Functions Equations and their Applications", Academic Press, New York.
33. R.S. Anderssen and F.R. de Hoog (eds) "The Applications of Mathematics in Industry", Wijhoff, Australia.
34. M.R. Ball "Mathematics in the Social and Life Sciences", Ellis Horwood and John Wiley.
35. R. Bellman and S.E. Dreyfus "Applied Dynamic Programming", Princeton University Press.
36. J.S. Berry, D.N. Burghes, I.D. Huntley, D.J.G. James and A.O. Moscardini, "Teaching and Applying Mathematical Modelling", Ellis Horwood and John Wiley.
37. J.S. Berry, D.N. Burghes, I.D. Huntley, D.J.G. James, and A.O. Moscardini, "Mathematical Modelling: Methodology, Models and Micros", Ellis Horwood and John Wiley.
38. J.S. Berry, D.N. Burghes, I.D. Huntley, D.J.G. James and A.O. Moscardini, "Mathematical Modelling Courses", Ellis Horwood and John Wiley.
39. W. Boyce (ed) "Case Structures in Mathematical Modelling", Pitmans, London.
40. F. Brauer and J.A. Nohal "Ordinary Differential Equations", N.A. Bejamins, New York.
41. F.S. Budnick "Applied Mathematics for Business, Economics and Social Sciences", McGraw Hill, New York.
42. D.N. Burghes "Mathematical Modelling in the Social Management and Life Sciences", Ellis Horwood and John Wiley.
43. D.N. Burghes and A.D. Wood "Mathematical Models in Social, Management and Life Science", Ellis Horwood and John Wiley.
44. D.N. Burghes, I.D. Huntley and J. Macdonald" Applying Mathematics" Ellis Horwood and John Wiley.
45. D.N. Burghes "Modelling with Differential Equations", Ellis Horwood and John Wiley.
46. H. Burkhardt "The Real World and Mathematics", Blackie.
47. F. Chorlton "Ordinary Differential and Difference Equations" Von Nostrand, New York.
48. C. Clark "Mathematical Bioeconomics", John Wiley.
49. P. Costello, D. Jones and B. Philips "Mathematics and Manufacturing" The Institute of Engineers, Australia.
50. R.A. Coddington and N. Levinson "Theory of Ordinary Differential Equations", Tata McGraw-Hill, New Delhi.

51. **M.** Cross and A.O. Moscardini, "The Art of Mathematical Modelling", Ellis Horwood and John Wiley.
52. C. Dyson, E. Ivery, "Principle of Mathematical Modelling", Academic Press, New York.
53. EDC/UMAP "Undergraduate Mathematics and its Applications", Project Publications, EDC. Camb. Mass.
54. L. Elsgotts "Differential Equations and Calculus of Variations", Mir Publishers, Moscow.
55. G.N. Ewing, "Calculus of Variations with Applications", McGraw Hill, New York.
56. F.R. Giordano and M.D. Weir "A First Course in Mathematical Modelling", Brooks Cole, California.
57. G. Hadley "Linear Programming", Addison Wesley, New York.
58. F. Harrary, "A Seminar on Graph Theory" Holt Rinehart and Winston, New York.
59. F.B. Hilderbrand "Advanced Calculus with Applications", Prentice Hall, New York.
60. F.B. Hilderbrand "Methods of Applied Mathematics", Prentice Hall, New York.
61. A.G. Howson and R. Mclone "Mathematics at Work", Heinemann, London.
62. J.D. Huntley and D.I.G. James (eds) "Case Studies in Mathematical Modelling", Oxford University Press.
63. D.M. Ingils "Computer Models and Simulation", Marcell Dekker.
64. J. Irving and M. Milleux "Mathematics in Physics and Engineering" Academic Press, New York.
65. S. Jacoby and J. Kowalik "Mathematical Modelling with Computers" Prentice Hall, New York.
66. D.J.G. James and J.J. Macdonald (eds) "Case Studies in Mathematical Modelling", Stanley Thames. Cheltonham.
67. J.N. Kapur "Insight into Mathematical Modelling", Indian National Science Academy, New Delhi.
68. J.N. Kapur "Maximum Entropy Models in Science and Engineering", Wiley Eastern, New Delhi and John Wiley, New York.
69. J.N. Kapur "Mathematical Models of Environment", Indian National Science Academy, New Delhi.
70. J.N. Kapur and H.K. Kesavan "Generalised Maximum Entropy Principle", Sandford Educational Press, Waterloo, Canada.
71. E. Kreyszig "Advanced Engineering Mathematics" Wiley International Edition.
72. R. Lesh, M. Nass, D. Lee, "Applications and Modelling" In Proc. of the Fifth Int. Cong. Mathematics Education, Birkhauser.
73. J. Medhi "Stochastic Processes", Wiley Eastern, New Delhi.
74. P.M. Morse and H. Feshback "Methods of Mathematical Physics", Vols I-II McGraw Hill, New York.
75. F.B. Murnagham "Introduction to Applied Mathematics", John Wiley, New York.
76. A.D. Myskis, "Introductory Mathematics for Engineers", Mir Publishers, Moscow.
77. F. Olivers and Pinto "Simulation Concepts in Mathematical Modelling", Ellis Horwood and John Wiley.
78. M.R. Osborne and R. Watts "Simulation and Modelling", University of Queensland Press, Brisbane.
79. Open University, UK, "Modelling Mathematics".
80. Open University, UK "Mathematical Models and Methods"
81. O. Ore "Graphs and their Uses", Random House.
82. L.S. Pontryagin et al. "The Mathematical Theory of Optimal Processes", Inter-Science, New York.

83. K. Reklony "Survey of Applicable Mathematics", MIT Press, Camb. Mass.
84. M.G. Smith "Theory of Partial Differential Equations". D Von Nostrand, New York.
85. I.N. Sneddon "Uses of Integral Transforms", Tata McGraw Hill, New Delhi.
86. I.N. Sneddon "Elements of Partial Differential Equations", McGraw-Hill, New York.
87. I.S. Sokolinokoff and R.M. Radheffer "Methods of Physics and Modern Engineering" McGraw Hill, New York.
88. I.S. and E.S. Sokolinkoff "Higher Mathematics for Engineers and Physicists", McGraw Hill, New York.
89. A.J.M. Spannier et al. "Engineering Mathematics, Vols. I-II" EOBC, London.
90. C. Seshu and M.B. Reed "Linear Graphs and Electrical Networks" Addison Wesley, New York.
91. D.K. Sinha and A. Misra, "Studies in Environmental Mathematics", South Asia Publishers, New Delhi.
92. T. Saaty and J. Alexander "Thinking with Models", (Mathematical Models in Physical, Biological and Social Sciences) Pergamon Press, New York.
93. The Spode Group "Solving Real Problems with Mathematics", Craneford Press.
94. M. Tenenbaum and H. Pollard "Ordinary Differential Equations", Harper and Row, New York.
95. Unesco "Applications of Mathematics", Chapter VII of New Trends in Mathematics Teaching III, Paris.
96. J.A. Trerney "Differential Equations", Allen and Backeon, New York.
97. H. Waylend "Differential Equations in Science and Engineering" Von Nostrand, New York and East West Press, New Delhi.
98. C.R. Wylie "Advanced Engineering Mathematics", McGraw-Hill, New York.
99. R. Weinstock "Calculus of Variations with Applications", McGraw-Hill, New York.
100. D.J. White "Dynamic Programming", Addison Wesley, New York.
101. B.P. Zeigler "Theory of Modelling and Simulation" John Wiley, New York.

Index

Actuarial science 108-109
Age-structured population models 58, 113
Allen's speculative model 65
Allocation problem 216, 218
Antibalance of a graph 163
Arms race model 70
Assignment model 229

Balance of a signed graph 161
Bipartite graph 168
Birth-death emigration-immigration
    processes 139
Bivariate normal distribution 207
Bose-Einstein distribution 242
Boundary conditions 147
Boundary-value problems 149, 188, 190
Brachistrochrone problem 210
Business models 26

Calculus of variations 204, 205, 219
Cargo-loading problem 216
Catenary 91, 210
Central forces 76, 78
Chaos 113
Chemical reactions 37
Circular motion 82
Cliques detection 159
Cobweb model 106
Communication network 157, 167
Compartment models 39-42, 63
Competition models 55
Compound interest 33, 173
Computerised tomography 247
Continuity equation
    fluid dynamics 126
    heat flow 127
    traffic 128
Curves of pursuit 94

Defective-coin problem 221
Degree of unbalance of a graph 163
Delay-differential equations 194
Diabetes mellitus 69

Difference equations
    linear 98-102
    non-linear 103
    stability of solution 104
    in population dynamics 110
    in genetics 114
Difference equation models 96-97
Differential-difference equations 194, 195
Diffusion 34, 40
Diffusion equation 129
Dirichlet distribution 207
Dissolution 37
Distance
    of moon 20
    of stars 21
Domar-debt models 65
Domar-Macro models 64
Duobalance of a graph 163
Dynamic models 72
Dynamic programming 181, 204, 214, 219,
    221

Economic models 64
Economo-dynamics 245
Elasticity 184
Electrical circuits 89
Electrical network 170
Elliptic partial differential equations 145
Elliptic orbits 15
Elliptic sound gallery 14
Emigration 32
Entropy 19, 179
Epicycloids 14
Epidemic models 60-62
    stochastic 140
Euler-Lagrange's equation 136, 205
Euler's formula for polygonal graphs 173
External ballistics 74

Farmer's problem 218
Feature extraction 203
Fermat's principle 15

Fermi-Dirac distribution 243
Fluid dynamics 126, 146, 211
Food webs 157
Forest management 227
Functional equations 177, 179, 180, 181

Gambler's ruin problem 119
Gamma distribution 206
Genetics models 114
Geodesics 218
Geometrical models 14, 48
Growth of science 32
Graphs
    bipartite 118
    complete 152
    directed 152
    genetic 155
    planar 170
    polygonal 173
    signal flow 168
    signed 153
    weighted 153
Gravity model 241

Hamilton's principle 202, 211
Harrary's measure 14
Harrod model 115
Heat flow 127
Hyperbolic partial differential equations
    145

Immigration 32
Infectious diseases 36
Information theory 179, 180, 234, 235
Initial conditions 147
Integral equations 184-192
    in elasticity 180
    from differential equations 187
    for two point problems 188
    in population dynamics 190
    in mathematical modelling 192
Integral transforms 185
Integro-differential equations 192
International trade model 72
Intermediate statistics distribution 244
Isoperimetric problem 209
Inventory control 24, 221

Kepler's laws 15, 80
Kirchhoff's laws 170

Map-colouring problems 172
Markov chains 117, 105
Mass-balance equations 126

Mathematical modelling through
    Algebra 16-20
    Calculus 23-28
    Calculus of variations 201-214
    Delay-differential and Integro-differential Equations 194-200
    Difference equations 96-123
    Dynamic programming 214-224
    Functional equations 177-184
    Geometry 14-16
    Graphs 151-176
    Integral equations 184-194
    Mathematical programming 225-237
    Maximum principle 237-240
    Maximum entropy principle 241-248
    ODE of first order 30-52
    Partial differential equations 126-150
    Systems of ODE's of first order 53-75
    Trigonometry 20-21
Mathematical bioeconomics 212
Mathematical programming 204
Matrices associated with a graph 150
Maximum entropy distributions 205
Maximum entropy principle 203, 234, 240-248
Maximum principle 204, 237, 240
Maximum likelihood estimation 181, 202
Maxwell-Boltzmann distribution 240
Minimum ehi-square principle 202
Minimum discrimination information principle 203, 234
Minimal surfaces 137, 208
Momentum-balance equation 132
Multispecies models 56, 195
Multi-stage rocket 47

Nature of partial differential equations 145
Non-linear programming 233-236
Normal distribution 206

Optics 213
Optimal exploitation of fisheries 222
Optimal harvesting of animal populations 239
Optimal portfolio selection 232
Optimization principles 201
Orthogonal 69

Parabolic partial differential equations 145
Parabolic mirrors 14
Pareto optimality principle 203
Partial differential equation models 124
Phillips stabilization model 91
Planar graphs 171

Planetory motion 76
Pollution effect 196
Pollution control 35
Population growth models 30, 53, 110, 427, 114
Portfolio analysis 203
Prey-predator models 53, 194, 177, 196
Price change 34
Principle of
    least time 202
    least action 202
    maximum entropy 203
    maximum likelihood 203
    minimum chi-square 203
    minimum potential energy 202
    optimality 203
Purchase problem 218

Quadratic programming 232

Radial velocity and acceleration 77
Radioactive decay 33
Radius of earth 16
Rectilinear motion 38
Regular solids 174
Reliability of devices 217
Resisting medium 45
Richardson's model 70
Rocket motion 45
Rotating fluids 211

Samuelson's interaction models 107-108
Samuelson's investment models 66-77

Satellite motion 85-86
Shannon's inequality 19
Shortest distance 207-208
Senior-subordinate relationship 155
Seven bridges problem 151
Signal flow graph
Simple harmonic motion 43
Stability of a market 67
Stability of equilibrium position 195
Stability of fixed points 112
Stability of prey-predator model 197
Statistical machanics 220-245
Stochastic epidemics model 140
Structure theorem for a graph 162

Technology innovation 36
Temperature change 34
Time-optimal problems 238
Tournaments 154
Traffic flow 128, 141-144, 155
Transportation problem 219, 229, 246
Transverse velocity and acceleration 77

Unoriented graphs 70
Urban and regional planning 246

Variational principles 36
Vibrating string 133, 137
Vibrating membrane 134, 138

Wave equation 134-136
Weighted digraphs 164, 165, 167

Mathematical Modelling

Each chapter of the book deals with mathematical modelling through one or more specified techniques. Thus there are chapters on mathematical modelling through algebra, geometry, trigonometry and calculus, through ordinary differential equations of first and second order, through systems of differential equations, through difference equations, through partial differential equations, through functional equations and integral equations, through delay-differential, differential-difference and integro-differential equations, through calculus of variations and dynamic programming, through graphs, through mathematical programming, maximum principle and maximum entropy principle.

Each chapter contains mathematical models from physical, biological, social, management sciences and engineering and technology and illustrates unity in diversity of mathematical sciences.

The book contains plenty of exercises in mathematical modelling and is aimed to give a panoramic view of applications of modelling in all fields of knowledge. It contains both probabilistic and deterministic models.

The book presumes only the knowledge of undergraduate mathematics and can be used as a text book at senior undergraduate or post-graduate level for a one or two-semester course for students of mathematics, statistics, physical, social and biological sciences and engineering. It can also be useful for all users of mathematics and for all mathematical modellers.